Searching for Dark Matter with Imaging Atmospheric Cherenkov Telescopes

Alessandro Montanari · Emmanuel Moulin

Searching for Dark Matter with Imaging Atmospheric Cherenkov Telescopes

 Springer

Alessandro Montanari
ZAH
Heidelberg University, Landessternwarte
Heidelberg, Germany

Emmanuel Moulin
Irfu/DPhP
CEA, Université Paris-Saclay
Gif-sur-Yvette, France

ISBN 978-3-031-66469-4   ISBN 978-3-031-66470-0 (eBook)
https://doi.org/10.1007/978-3-031-66470-0

This Springer imprint is published by the registered company Springer Nature Switzerland AG
The registered company address is: Gewerbestrasse 11, 6330 Cham, Switzerland

If disposing of this product, please recycle the paper.

"Considerate la vostra semenza:
fatti non foste a viver come bruti,
ma per seguire virtute e canoscenza"

Dante, Divina Commedia, "Inferno", canto
XXVI, vv. 118–120

Preface

Very-high-energy (VHE, $E \underset{\sim}{>} 100$ GeV) photons are a powerful probe for studying astrophysics and fundamental physics under extreme conditions. During recent years, the research field of gamma-ray astronomy brought a continuous increase in the knowledge of the most violent phenomena in the Universe via the detection of copious amounts of VHE gamma rays in a variety of environments. The advent of new and more powerful detectors, designed to detect the short Cherenkov light pulse produced in the air showers initiated by VHE gamma rays interacting with the Earth's atmosphere, enabled to tremendously go well beyond initial discoveries.

VHE gamma rays are also efficiently exploited to probe fundamental physics such as the nature of Dark Matter (DM). This elusive component, required to address several open questions in cosmology and astrophysics, could be entirely made of Weakly Interacting Massive Particles (WIMPs). Imaging Atmospheric Cherenkov telescopes (IACT), such as the High Energy Stereoscopic System (H.E.S.S.), could observe gamma-ray products in the final state of WIMP self-annihilation. A wealth of constraints on dark matter particle properties has been produced by IACTs.

A number of reports have been published on dark matter searches with gamma rays. It is now relevant to present a review on this topic for graduate students with astrophysics and particle physics backgrounds in light of the recent achievements in the VHE astrophysics field and the new observatories to come by the end of the decade. We present current achievements, the technique at the base of the functioning of IACTs, the framework developed to search for DM particle models and distribution, and how to use the collected datasets to obtain meaningful insights on the particle nature of DM. We also provide detailed examinations of the results obtained so far with IACT observations on the search for DM annihilation signals, together with outlooks on the expectations for future observations in this field, which is timely given the near-future advent of the Cherenkov Telescope Array.

Alessandro Montanari (AM) is indebted to A. De Angelis, University of Padova, who thoroughly and extensively reviewed his Ph.D. thesis. AM also wishes to thank Emmanuel Moulin (EM), who initiated him into the field of gamma-ray astronomy and DM search. AM and EM were encouraged to assemble all the ingredients for a

DM signal search using IACT observations into this book so that it may serve future generations of astronomers working in this lively field. AM work has been funded by the Deutsche Forschungsgemeinschaft (DFG, German Research Foundation)— 460248186. EM wishes to acknowledge the hospitality of the University of Adelaide, where part of the writing was performed.

Heidelberg, Germany   Alessandro Montanari
Gif-sur-Yvette, France   Emmanuel Moulin
March 2024

Contents

List of Figures

List of Tables

Chapter 1
Introduction

The study of non-thermal processes in the Universe, the most violent ones, has been pioneered by Very High Energy (VHE, E $\gtrsim$ 100 GeV) astrophysics. VHE gamma-rays serve also for delving into fundamental physics beyond the standard model of particle physics. The major portion of the matter content of the universe, approximately 85%, is composed of Dark Matter (DM), the enigmatic nature of which continues to confound scientists. One of the leading explanations for DM involves non-baryonic massive particles undergoing, beyond gravitational interaction, weak force interaction with standard matter. These are termed weakly interacting massive particles (WIMPs). If WIMPs are massive enough, they can undergo self-annihilation generating TeV gamma-rays in dense regions of the cosmos. The Galactic Center (GC), along with its immediate surroundings, stands out as a prime location for detecting signals from DM annihilation.

Imaging Atmospheric Cherenkov Telescopes (IACT), such as the High Energy Stereoscopic System (H.E.S.S.)—a Southern-hemisphere array of 5 IACTs, can detect VHE gamma-rays in the $\sim$ 100 GeV to $\sim$ 100 TeV energy range. From around the years' 2000, VHE astrophysics has evolved, producing more and more forefront results. Cosmic rays, accelerated by various galactic and extragalactic phenomena in environments such as supernova remnants, black holes, and active galactic nuclei, emit gamma-rays, photons with energies surpassing the MeV threshold. Despite extensive research, the mechanisms driving the origin and acceleration of cosmic rays remain subjects of ongoing lively debate. The GC stands out as a particularly promising area in the local Universe for studying these phenomena, as well as for probing potential signals from DM.

The GC region offers a prime target for observation with the H.E.S.S. instrument due to its advantageous location in the Southern hemisphere, enabling the highest sensitivity among currently operating IACTs. This particular region of the sky holds significant promise for observations, especially concerning potential signals

A. Montanari and E. Moulin, *Searching for Dark Matter with Imaging Atmospheric Cherenkov Telescopes*, https://doi.org/10.1007/978-3-031-66470-0_1

of DM annihilations within mass ranges that cannot be effectively explored at collider experiments. Thanks to observations of the GC region, strong constraints have been established on DM and emissions at very high energies. Looking ahead, the Cherenkov Telescope Array (CTA), representing the next generation of IACTs, is poised to enhance sensitivity in the TeV energy range further, facilitating deeper investigations into these emissions.

Since the GC is one of the most promising laboratories to study possible DM annihilation signals in the VHE range, and H.E.S.S. is the best-placed instrument observing the GC region at the moment of the writing, this book is dedicated to exploring what the current sensitivity reach to DM annihilation signals with IACT observations is. H.E.S.S. is recalled as an instrument sufficiently representative of the telescope class to investigate the topics. The extensive program of observations of the Galactic Center region with H.E.S.S., the Inner Galaxy Survey, will be described. This dataset is used to investigate diffuse emissions in the region and search for Dark Matter annihilation signals from the inner halo of the Milky Way.

The book's first part aims to introduce the Dark Matter paradigm and how to potentially detect signals from the annihilation of Dark Matter particles indirectly. In this part, the pieces of evidence for the Dark Matter existence, the candidates to explain this elusive component—with a focus on particle candidates and specifically on WIMPs, some theoretical bounds on these particle candidates, and the detection techniques that are being exploited to grasp a DM signal are reviewed. Then, the focus is placed on the indirect detection technique, for which the main ingredients are discussed: how one should parameterize the expected gamma-ray flux from DM annihilation, the DM distribution in the observed target, and the adopted DM profiles.

The book's second part is dedicated to describing astrophysical phenomena that can be deeply investigated with IACTs. The non-thermal processes accelerating cosmic rays in our Universe and how gamma rays at TeV energies can be produced are first reviewed. The main experiments observing gamma rays at high- or very-high energies are then briefly outlined. The main characteristics of sources emitting TeV photons in the GC are also succinctly presented, being this the region of the sky that will be used for the analyses presented in the book's third part. These sources must be considered background emissions when looking for a faint DM annihilation signal. The technique at the base of the IACT functioning is then explained, with a review of the development of particle showers in the Earth's atmosphere, how these produce the Cherenkov light and how this one can be detected by H.E.S.S. Some more information about the H.E.S.S. array is outlined, together with its calibration, trigger system, event identification and reconstruction and performances. This part is concluded with a brief outlook on the expected performances of the future of TeV observatories, i.e., the Cherenkov Telescope Array.

The third and last part of the book shows the analyses and results of the indirect search for DM signals. This part starts with a chapter entirely dedicated to the statistical methods applied when searching for a faint emission—such as a DM signal—in regions of the sky crowded with background astrophysical sources. All the ingredients necessary to perform such kind of analysis are detailed: from the construction of a mock dataset to the definition of likelihood functions and test statistics, and finally, how to derive upper limits on free parameters of the model for the searched emission, including systematic uncertainties, Monte-Carlo realizations and performance tests when injecting a fake signal. Then, a real analysis with the Inner Galaxy Survey data collected by H.E.S.S towards the GC region is presented. Here, all the required steps, starting from the low-level analysis of the data, exposure maps, and instrumental systematic uncertainties, are explained. Then, the region of interest for the search for a DM signal and how to measure the residual irreducible background that affects IACT observations are defined. Once the expected DM annihilation signal is presented, and no significant excess compatible with DM is found in the dataset, the procedure for deriving upper limits is shown. These are obtained to constrain the annihilation cross section $\langle \sigma v \rangle$ of the DM particles for the assumed models. This part is concluded with a discussion on the sensitivity reach of the current generation of IACTs to TeV DM annihilation signals. Two publicly available tools to derive the expected gamma-ray yield from DM are presented together with some TeV DM candidates in the form of specific WIMP scenarios and alternative assumptions of the DM distribution in the GC region. For this alternative analysis, a different approach is implemented, where all the measurements and the possible sources of background in the GC for DM searches are modeled. Once this has been defined, the sensitivity reach is inspected, simulating observations for realistic live times at the moment of the writing. The last steps of this part are to discuss the main uncertainties affecting this limits computation—being them theoretical, from background modeling, or instrumental. An outlook on expectations from the future in this field of research is provided in the conclusive part of the book.

Part I
Dark Matter and Its Indirect Detection

Chapter 2
The Dark Matter Mystery

Abstract A substantial body of cosmological and astrophysical evidence points toward the existence of dark matter (DM) in our Universe. Nevertheless, the nature of this elusive component is still a mystery. This chapter is dedicated to an overview of the ΛCDM paradigm to explain particle dark matter. According to cosmological measurements, 24% of our Universe is constituted by non-baryonic DM. Adopting the DM paradigm can also explain measurements of astrophysical phenomena at the galactic scale. In this chapter, we first outline the pieces of evidence from astrophysics and cosmology for the existence of DM, and their potential issues at small scales with some alternatives. Then, we introduce widely motivated candidates to explain dark matter. We discuss the thermal freeze-out of Weakly Interacting Massive Particles DM and its expected thermal relic density together. Some existing bounds on particle dark matter models are also briefly discussed. We conclude the chapter by giving an overview of the nowadays available detection techniques that are deployed to shed light on the DM mystery.

Keywords Λ-CDM cosmology · Dark matter · Thermal relic density · WIMP miracle · Dark matter bounds · Axion-like particles · WIMPs · Primordial black holes · Detection techniques

2.1 Observational Evidence for Dark Matter

2.1.1 Evidence from Astrophysics

When it was understood that the measured velocity of objects in gravitationally bound systems was diverging from the expected one from the gravitational interaction with the visible matter, the first historical evidence of the necessity of dark matter (DM) to explain the standard model of Cosmology was ready. The measurements were taken by Fritz Zwicky, who computed the velocity dispersion of individual galaxies in the Coma cluster in the '30s (Zwicky 1937).

The virial theorem provides a relationship between the total mass M of a spherical system of radius R at equilibrium and the velocity dispersion σ_v of galaxies as

© The Author(s), under exclusive license to Springer Nature Switzerland AG 2024
A. Montanari and E. Moulin, *Searching for Dark Matter with Imaging Atmospheric Cherenkov Telescopes*, https://doi.org/10.1007/978-3-031-66470-0_2

$M(< R) = 5R\sigma_v^2/3G$ for an homogeneous sphere. Zwicky estimated about 800 galaxies in the Coma cluster, each with a stellar mass of $10^9 M_\odot$. With his estimate, $R \simeq 3$ Mpc and measurements of the radial velocities of the galaxies $\langle v_r^2 \rangle = (1000$ km/s$)^2$. Assuming equipartition of the kinetic energy $\langle v^2 \rangle = 3 \langle v_r^2 \rangle$ and applying the virial theorem, one obtains that the total mass is $\sim 4 \times 10^{14}$ $M_\odot$, about 400 times larger than the observed mass in galaxies. While we know today that a large fraction of the mass of the galaxy clusters is made of hot gas, this finding is considered a pioneer in discovering the missing mass in the universe. The gas is ionized and produces photons in X-rays via Bremsstrahlung that escapes the cluster. The X-ray flux can be computed using the production cross section of X-rays via Bremsstrahlung of non-relativistic electrons. It depends on the temperature and the density of the gas. For Coma, one can show that the mass of the gas is about 2×10^{14} $M_\odot$, which represents about 10% of the total mass of Coma. More than 85% of the mass in Coma is made of dark matter. Zwicky noted that according to the measure of visible mass, single galaxies in the Coma cluster were moving too fast for the latter to remain bound together. He posited that an unobserved type of mass—the *dunkle Materie* (Dark Matter)—might explain this.

Later in the '70s, another measurement of **galaxies' rotational curves** by Vera Rubin and Kent Ford also confirmed this hypothesized missing component (Rubin and Ford 1970). They found that the velocity of the stars in the Andromeda galaxy does not follow Kepler's law $1/\sqrt{r}$ behavior. Indeed, the velocity profile stayed constant in the outer galaxy. This directly implies additional invisible matter if Newtonian gravity is considered valid. The former would be a DM halo extending as $1/r^2$ from the galaxy's center, as expected in the isothermal sphere model (Binney and Tremaine 2008). Assuming hydrostatic equilibrium, a perfect gas, and spherical symmetry, one can show that $\sigma_v^2 d\rho/dr = -\rho GM/r^2$. For a Maxwell-Boltzmann distribution $f(E)$ with $E = p^2/2m + m\phi$, one can solve the Poisson equation $\Delta\phi = 4\pi G\rho(\phi)$ with $\rho(\phi) = \int d\vec{p} f(\phi)$ to obtain $\rho(r) = \sigma_v^2/2\pi G r^2$. In this case, mass diverges, so the non-singular isothermal sphere has then been proposed (Binney and Tremaine 2008).

Galaxy rotation curves are obtained with data from the stellar population in the inner part and measurements of the Doppler shift of the 21-cm emission line of neutral hydrogen in the outer one. This latter can cover faint regions beyond the disk at several tenths of kpc. Figure 2.1 shows the rotation curve of the galaxy NGC 3189, as a function of the distance from the galaxy's center (van Albada et al. 1985). The curve "disk" shows the contribution from only visible matter. What is expected from the DM halo is shown from the "halo" curve.

A shred of additional evidence from astrophysics' probes comes from **gravitational lensing** (Massey et al. 2010). The latter implies distortion of background light caused by the deformation of space-time due to gravitating mass. This deformation produces a lens effect on background galaxies, similar to optical refraction. This effect is shown in the left panel of Fig. 2.2, in the observations of the Abell 1689 cluster with Hubble (2010). The visible mass generates a potential that is not enough to explain the significant bend in the light coming from behind.

Among the most convincing evidence for DM existence are collisionless DM halos in the cluster merger E0657-558. This episode is famous under the name of the **Bullet**

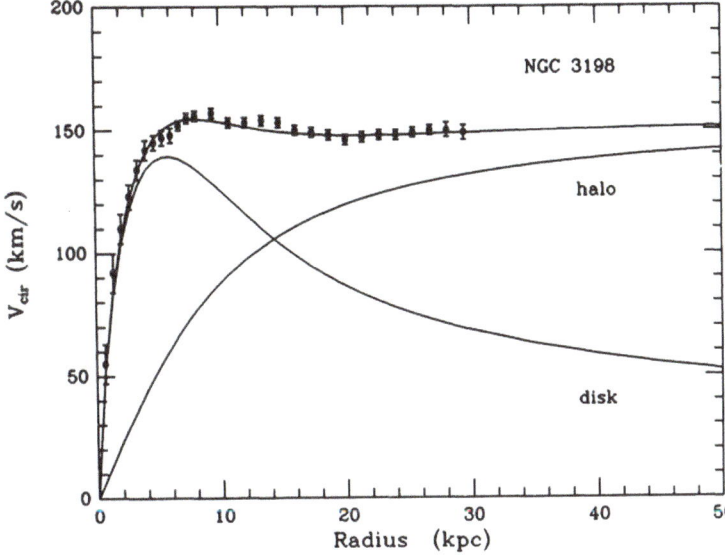

Fig. 2.1 Rotation curve of NGC 3198 galaxy. Dots represent measured points, shown together with the contribution from the visible matter ("disk" curve) and DM halo ("halo" curve). Figure extracted from van Albada et al. (1985)

Fig. 2.2 *Left panel*: Gravitational lensing of the Abell 1689 galaxy cluster observed with the Hubble telescope. The DM halo of the cluster distorts the light coming from the galaxies in the background. Figure extracted from New Hubble image of galaxy cluster (Abell 1689 (2010)). *Right panel*: The Bullet Cluster obtained as a composite image of the merger 1E0657-558. Credit: X-ray: NASA/CXC/CfA/ (Markevitch 2005); Lensing Map: NASA/STScI; ESO WFI; Magellan/U.Arizona/ Clowe et al. (2006); Optical: NASA/STScI; Magellan/U.Arizona/ (Clowe et al. 2006)

Cluster (Clowe et al. 2006; Markevitch 2005). The right panel of Fig. 2.2 shows the Bullet Cluster's composite X-ray and optical image. The magenta represents ordinary matter. The weak lensing of light passing close to a massive object is used to estimate the mass distribution, shown in blue. The latter is dominated by DM. The hot gas of the mergers lingers behind the subcluster galaxies and interacts; however, the DM component is ahead of the collisional gas and coincident as if the galaxies were collisionless. Measuring the Bullet Cluster has been used for constraining the self-interaction cross-section of DM down to $\sigma/m < 1$ cm^{2}g^{-1} (Markevitch et al. 2004).

2.1.2 Evidence from Cosmology

The standard model of Cosmology is founded on Einstein's General Relativity equations, Friedmann-Lemaître-Robertson-Walker metric, and the discovery of the expansion of the Universe (Kermack and McCrea 1933; Friedmann 1922; Lemaitre 1927; Robertson 1936).

  With this model, one can explain the thermal history from the Big Bang, the relic background radiation, the abundance of the elements, and the formation of structures at large scale. For extended reviews of the model, the interested reader can consult (Peebles 1980; Bertone et al. 2005).

  The cosmological principle states that, at a sufficiently large scale, the Universe is homogeneous and isotropic. The distribution of galaxies confirms homogeneity. The observations and measurements of Cosmic Microwave Background (CMB) can explain isotropy.

  To derive the Friedmann equations (Friedmann 1922; Robertson 1936; Lemaitre 1927), the already mentioned hypotheses are considered together with the relationship between the energy content and the geometry of the Universe (Einstein 1916). The equations are

$$\left(\frac{\dot{a}}{a}\right)^2 = \frac{8\pi G}{3}\rho - \frac{k}{a^2}$$
$$\frac{\ddot{a}}{a} = -\frac{4\pi G}{3}\left(\rho + \frac{3p}{c^2}\right) + \frac{\Lambda c^2}{3}. \tag{2.1}$$

The Newtonian gravitational constant is given by the term G. The curvature of the Universe is encapsulated in k, which can be -1 for an open hyperbolic space, 0 for a flat space, and $+1$ for a closed spherical space. The way the Universe expands is specified by the scale factor $a(t)$, and the vacuum energy that empowers the accelerated Universe expansion is given by Λ—also known as the cosmological constant. The sum of the energy densities of the Universe—including matter and radiation—is expressed by $\rho = \rho_m + \rho_r$. The pressure is the term p. The Hubble parameter H in the first equation, is given by $H(t) = \dot{a}(t)/a(t)$. If a flat Universe is assumed, the critical density $\rho_c = 3H^2/8\pi G$ is equaled by the total density ρ_{tot}. The

Table 2.1 Values of the cosmological parameters from Planck measurements presented in Planck 2018 results (2020)

Parameter	Symbol	Value
Hubble constant	$H_0 = 100\,h$ [km s^{-1} Mpc^{-1}]	67.4 ± 0.5
Cold DM density	$\Omega_{CDM}h^2$	0.120 ± 0.001
Baryon density	$\Omega_b h^2$	0.0224 ± 0.0001
Matter density	$\Omega_m = \Omega_b + \Omega_\chi$	0.315 ± 0.007
Curvature	Ω_k	0.001 ± 0.002
Vacuum energy density	$\Omega_\Lambda h^2$	0.3107 ± 0.0082
Cosmological constant	Λ [eV^2]	$(4.24 \pm 0.11) \times 10^{-66}$

latter includes the density of matter and radiation plus the density of the vacuum, *i.e.* $\rho_{tot} = \rho + \rho_\Lambda$, with $\rho_\Lambda = \Lambda/8\pi G$. Each component can be expressed as a fraction of the critical density in terms of a density parameter $\Omega_i = \rho_i/\rho_c$.

Rearranging the first Friedmann equation with the present values of the density parameters, *i.e.* using the relic density of matter, radiation, and vacuum energy, one obtains:

$$\frac{H^2(z)}{H_0^2} = \Omega_r(1+z)^4 + \Omega_m(1+z)^3 + \Omega_k(1+z)^2\Omega_\Lambda, \qquad (2.2)$$

where also the curvature term is given by $\Omega_k = -k/H_0^2$. The scale factor is related to the redshift in $a(t) = 1/(1+z)$. Cosmological probes can be used to measure that most of the matter is not made of baryons but of cold DM, which can be rendered through: $\Omega_m = \Omega_b + \Omega_\chi$, with Ω_χ being much larger than Ω_b. The DM is considered as a particle-like component. This widely accepted cosmological model, including dark energy and DM, is dubbed the Lambda-cold-dark-matter (ΛCDM) model.

The relic density values for each component in the ΛCDM model can be obtained from Planck measurements from Planck 2018 results (2020) and are shown in the Table 2.1.

The early Universe was permeated by a plasma of photons and baryons in thermal equilibrium, where free electrons could move. When the Universe's temperature cooled to $\sim 3,000$ K—at the recombination epoch—neutral hydrogen could form. Most of this primordially produced hydrogen was in excited states that transitioning to the bound state caused photons emission. Being the Universe transparent, the photons could propagate freely after this so-called decoupling era and constitute the fossil light of the Big Bang—the **Cosmic Microwave Background**. This was accidentally detected at the Bell Labs by a radio telescope in 1964 Penzias and Wilson (1965). Today's CMB measurements provide the relic temperature of the Universe, $T = 2.725$ K. In the '90s, anisotropies in the CMB were measured by Smoot et al. (1992) at the level of $16 \pm 4\,\mu$K. The direct correlation of the baryon density with the CMB temperature allows to derive the former. Cold spots for areas with high density, and warm ones for under-densities appear on the CMB map. Figure 2.3 shows the power

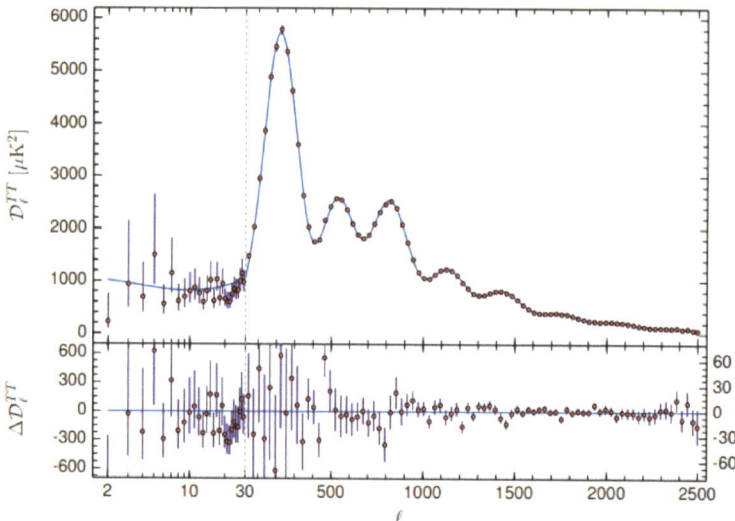

Fig. 2.3 CMB angular power spectra as a function of the multiple moments derived from Planck 2018 measurements. The ΛCDM predictions are fitted too. Figure extracted from Planck 2018 results (2020)

spectrum of CMB temperature as a function of the multipole moment. Radiation pressure and baryon gravitational potential provoke the compression peaks—shown as the odd ones. Decompression, which is driven only by radiation, produces the even peaks. A large baryon density would be reflected on larger odd peaks compared to even ones. The relative amplitude between the second and the first peaks provides then a measurement of Ω_b.

The anisotropies in the CMB, from the photon-baryon fluid detectable today, can be explained through the concept of **Baryonic Acoustic Oscillations** (BAO). The interplay between baryonic gravitational potential and radiation pressure formed the relativistic acoustic waves in the early Universe's primordial plasma. Traveling at the speed of sound, the baryon-photon fluid propagated outwards as an expanding shell from the overdensities' positions, powered by radiation pressure. When baryons were no more under radiation pressure's influence, the acoustic waves froze—at the recombination epoch. This implies that the spherical baryonic shells stood around the central DM overdensities. Similar to the comoving radius of the baryons' shell is the comoving sound horizon at recombination r_s. So, BAOs can be used as standard rulers for the horizon size and the geometry of the Universe. Driven by gravity, baryonic and DM density evolve together. A separation of r_s is more likely to be observed between nowadays observed galaxies. This is reflected in a peak at r_s in the density profile of the matter. The density profile is obtained from many perturbations, *i.e.* requires a statistical correlation between the position of large-scale structures in the Universe, and it is not measurable with one single object. With no DM, there would be no characteristic correlation scale due to the complete removal of the perturbations. The reader interested in a BAOs review can consult (Bassett and Hlozek 2009).

Type Ia Supernovae produce a fairly consistent and high luminosity because of the fixed value of the critical mass at which the explosion of the white dwarf occurs. These objects are not identical, though they constitute a type. They can be used as standard candles because their behavior depends on local physics and is expected to be independent of environment and evolution. Therefore, their distance is the factor influencing their apparent magnitude. The expansion of the Universe causes their redshift. Through measurements of these supernovae, $a(t)$ can be constrained, and one can then put constraints on the relic densities in the ΛCDM model (Peebles 1980).

The process of primordial nucleosynthesis, or **Big Bang Nucleosynthesis**, is of fundamental importance when determining the abundance of baryons (Coc and Vangioni 2017). It is responsible for creating the chemical elements during the early phases of the Universe after the Big Bang. While the Universe was still hot, the light elements were formed by nuclear reactions in the first tens of minutes. Widely accepted values for the abundance of these elements—^{4}He, D, ^{3}He and ^{7}Li—are fixed: ^{4}He/H ~ 0.1, ^{3}He/H $\sim$ D/H $\sim 10^{-5}$ and ^{7}Li/H $\sim 10^{-10}$. The ratio of baryon-to-photon η, constrained in the range of $5.1 \times 10^{-10}\eta < 6.5 \times 10^{-10}$, determines the abundance of light elements (The Baryonic Density 2010). The abundance of baryonic matter correlates to η and it is measured as $\Omega h^2 = 0.0224 \pm 0.0001$ (Planck 2018 results 2020).[1] It is shown to account for about 5% of the critical density of the Universe and to be five times smaller than DM abundance. The other 95% to complete Universe's density is constituted by 70% of dark energy and by 25% of DM. The nucleosynthesis of the baryons is one of the main proofs of the validity of the ΛCDM model (Deruelle and Uzan 2018).

The hierarchical **Structure Formation** is due to amplifying fluctuations in the primordial density growing because of the Universe's expansion (Primack 1997). The measurement of the distribution of luminous objects in the Universe can be used to describe the formation of large-scale structures and understand their relation to the characteristics of observed objects. How the matter is distributed in the sky can be explained via surveys combining measurements of the redshift and the angular position of astronomical objects. The 2dF Galaxy Redshift Survey (2dFGRS) conducted by the Australian Astronomical Observatory mapped a statistically representative volume of the Universe (The 2dF Galaxy Redshift Survey 2001). The survey revealed optically luminous galaxies in that volume. A more detailed three-dimensional map of a third of the sky was produced by the Sloan Digital Sky Survey (SDSS) (The Sloan Digital Sky Survey 2000), applying spectra produced with multi-color images in ultraviolet, green, red, and infrared. We can compare the observed structure distribution to the one simulated from the growth of the cosmic fluctuations in the near-uniform early Universe. The model cannot sustain analytical treatment

---

[1] A common practice is to introduce the dimensionless Hubble constant, usually denoted by h and commonly referred to as "little h" (Croton 2013). Then, the Hubble constant is written as $H_0 = h \times 100$ km s^{-1} Mpc^{-1}.

of gas dynamics, radiative cooling, photoionization, recombination, and radiative transfer. Thus, complex N-bodies numerical simulations in a large box of space are required. The Universe's initial conditions set nearly uniform matter density with small inhomogeneities. The simulation of perturbations follows the measured CMB temperature power spectrum. Then the equations describing the Universe's expansion, gravity, baryonic gas pressure forces, and dark energy are injected. Only recently, baryons' effects have been included in the simulations. However, consensus on how to include these physical processes is still lacking. The fluctuations develop with the evolution of the system due to gravity. High initial density regions form DM halos and galaxies because matter collapses here. Springel et al. (2005), Springel et al. (2006) can be consulted by the reader interested on more details on the simulations of the formation of large-scale structures, clusters, and groups of galaxies and their evolution. When filaments become more prominent, clusters can also form at the intersections between them. Structures' growth slows down at a redshift larger than 1 because gravity becomes subdominant and dark energy dominates the acceleration. First, low-mass objects are formed and later merged into bigger ones. Hot DM cannot explain the distribution of the nowadays observed galaxy-scale structure, as seen from simulations. Only cold DM was included in the setup of initial simulations of structure formation. Very cuspy profiles were predicted, even though these are in tension with observations at galactic scales. The inclusion of baryons tends to flatten the inner part of the halos. The left and top panels of Fig. 2.4 show part of the 2dFGRS and SDSS maps, respectively. Surveys obtained with the Millenium simulations are shown in the opposite panel (Formation of the large-scale structure in the Universe 2010). These utilize semi-analytic techniques for the simulation of the dark matter distribution and structure formation and the experimental surveys regarding geometry and magnitude limits. A non-baryonic DM component can be included in the matter content to obtain a striking agreement between the simulations and the measurements. Considering baryonic-only matter would not allow the fluctuations to reproduce the evolution and the formation of the observed structures from the early Universe to today. Prominent structures, like the observed Sloan Great Wall, are also obtained in the simulations (visible in the top panel of Fig. 2.4).

2.1.3 Thermal Relic Density of Cold Dark Matter Particles

Thermal DM production is a central assumption of the standard DM picture (see e.g. Gelmini and Gondolo (2010); Baer et al. (2015) for a discussion on DM production mechanisms). Here, DM particles are relics of the Big Bang: they were thermally produced in the early Universe, *i.e.* from particles in thermal equilibrium. Among the most popular thermal relics are massive particles coupled to Standard Model (SM) particles via the weak interaction—the Weakly Interacting Massive Particles (WIMPs, which are discussed more in detail later in Sect. 2.2.3), and no asymmetry is assumed between DM particles and antiparticles.

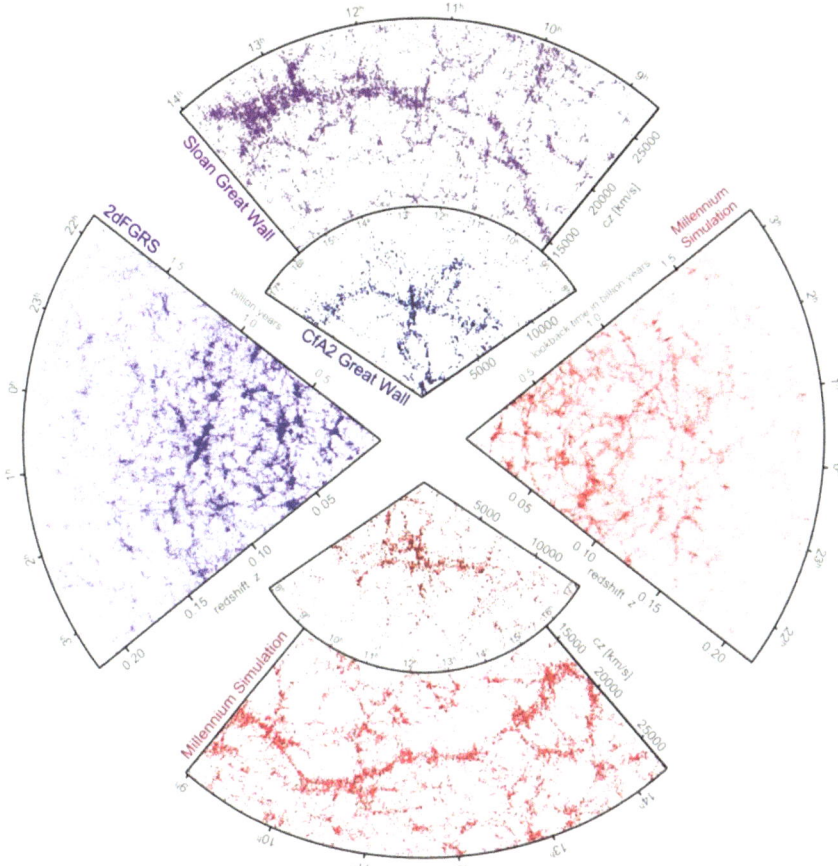

Fig. 2.4 The left panel shows the map of a part of the 2dF Galaxy Redshift Survey. The top one shows a part of the Sloan Digital Sky Survey. Maps obtained for the corresponding portions of the sky with the Millenium simulations are shown in the right and bottom panels. Figure extracted from Springel et al. (2006)

All the particles composing the early Universe are in a thermal bath (Bergstrom 2000) when the temperature T is much higher than their mass m. Among other particles, WIMPs can be produced and destroyed. The temperature T decreases, exponentially suppressing their density with m/T. The equilibrium is left when the temperature is not high enough to sustain the pair production of WIMPs, and only the annihilation process $\Gamma = n\sigma v$ among the particles, which depends on the velocity v and its cross-section σ, remains possible. As the universe expands, when the WIMP mean free path becomes comparable to the Hubble distance, WIMPs cannot self-annihilate anymore. Their co-moving density remains then constant; this process is commonly referred to as *freeze-out*. The remaining diluted abundance of WIMPs may constitute the DM today.

Different particle masses mean different decoupling epochs, because of their intrinsic properties and interaction processes.

Following (Kolb and Turner 1990), the evolution of n in thermal and chemical equilibrium in the early Universe is given by the Boltzmann equation as:

$$\frac{dn}{dt} = -3 Hn - \langle\sigma v\rangle(n^2 - n_{eq}^2) \tag{2.3}$$

where H stands for the Hubble parameter, $\langle\sigma v\rangle$ the thermal average of the annihilation cross-section of DM particles times relative velocity, and n_{eq} is the equilibrium number density of DM particles. The n^2 term arises from processes $DMDM \rightarrow SMSM$ that destroy DM particles, where SM denotes SM particles, and the n_{eq}^2 term arises from the reverse process $SMSM \rightarrow DMDM$, which creates DM particles. The dilution of the WIMPS due to cosmic expansion is represented by the first term on the right-hand side and the decrease of the number of particles by annihilation is given by the second term. The thermal relic density is determined by solving the Boltzmann equation numerically. A simplification of this equation makes use of the conservation of entropy. The entropy per comoving volume expresses as $s = 2\pi^2 g_* T^3/45$ (Kolb and Turner 1990). g_* is the number of relativistic degrees of freedom of the DM particles. Its dependence with temperature can be neglected in the following estimate given that it is a slowly varying function of the temperature. Using the conservation of entropy per comoving volume, $S = sa^3 = constant$, which implies $ds/dt = -3Hs$, and the change of variables $Y = n/s$, Eq. (2.3) becomes:

$$s\dot{Y} = -\langle\sigma v\rangle s^2(Y^2 - Y_{eq}^2) \tag{2.4}$$

where $\dot{Y} \equiv dY/dt$ and $Y_{eq} = n_{eq}/s$. In the regime of radiation-dominated era, the Hubble parameter reads $H = (4\pi^3 g_*/45)^{1/2} m_{DM}^2/(x^2 M_{Pl})$ where $x \equiv m/T$ and where M_{Pl} is the Planck mass.[2] In the radiation-dominated regime, $t = 1/2H$. Differentiating this expression leads to $dx/dt = Hx$ and Eq. (2.3) eventually takes the following simplified form:

$$\frac{dY}{dx} = -A\frac{\langle\sigma v\rangle}{x^2}(Y^2 - Y_{eq}^2) \tag{2.5}$$

with $A \equiv m_{DM}M_{Pl}(\pi g_*/45)^{1/2}$. This is a Ricatti-type equation that cannot be solved analytically. Nevertheless, the late-time value of the density can be obtained analytically. The equilibrium distribution is obtained by the late-time density and can be obtained analytically by $Y_{eq}(x) = 45g/(4\pi^4 g_*)x^2 K_2(x)$, where g is the number of internal degrees of freedom of the particle and K_2 the modified Bessel function

---

[2] For example, at a temperature of 1 TeV, all the particles of the Standard Model were relativistic and in thermal equilibrium. $g(\text{fermions}) = 6[u, d, c, s, b, t] \times 2[U(1)] \times 3 \times [SO(3)] \times 2(SU[2]) = 90$. $g(\text{bosons}) = 2[\gamma] + 3[W^{\pm}, Z] + 8[g] + 1[h]$. The total number of internal degrees of freedom of the fermions is 90 and for the gauge and Higgs bosons 28, so the total for g_* is $g_*(T = 1\text{TeV}) = 28 + 7/8 \times 90 = 106.75$.

of the second kind. In the non-relativistic regime corresponding to mT, due to the exponential decrease of Y_{eq} in the nonrelativistic regime corresponding to $x \gg 1$, Y_{eq}^2 can be neglected with respect to Y^2. In order to further investigate the dependency of the relic density with the annihilation cross section, a simple assumption is that the latter is independent of the velocity, i.e. independent of the temperature, which is known as the s-wave contribution. After the separation of variables and integration between the freeze-out x_F and $+\infty$ (T$\to$0), one obtains:

$$\frac{1}{Y_F} - \frac{1}{Y_0} = -A \frac{\langle \sigma v \rangle}{x} \tag{2.6}$$

With $Y_0 \ll Y_F$, the number density today is $Y_0 = x_F/(m_{DM} M_{Pl}(\pi g_*/45)^{1/2} \langle \sigma v \rangle)$. The relic abundance of DM particles is given by $\Omega_{DM} = \rho_{DM}^0/\rho_c$ where the critical density is $\rho_c = 1.05 \times 10^{-5} h^2$ GeV cm^{-3}. With $s_0 = m_{DM} n_0 = m_{DM} s_0 Y_0$ where $s_0 = 2889.2$ cm^{-3} is the present-day entropy, one gets:

$$\Omega_{DM} h^2 = 1.04 \times 10^9 x_F g_*^{-1/2} (M_{Pl}/\text{GeV})^{-1} (\langle \sigma v \rangle/\text{GeV}^{-2})^{-1} \tag{2.7}$$

In order to determine the relic abundance dependence with $\langle \sigma v \rangle$, one needs to estimate x_F. The freeze-out temperature is obtained when $n \langle \sigma v \rangle \equiv H$. Now, $n \simeq n_{eq}$, therefore using the equilibrium density in the non-relativistic regime:

$$n_{eq} = g \left(\frac{m_{DM} T}{2\pi} \right)^{3/2} \exp\left(-\frac{m_{DM}}{T} \right) \tag{2.8}$$

and the expression of H, one gets:

$$x_F = \ln\left(\frac{g \langle \sigma v \rangle m_{DM} M_{Pl}}{2^{5/2} \pi^3} \left(\frac{45}{g_*} \right)^{1/2} x_F^{1/2} \right). \tag{2.9}$$

x_F can be estimated by solving this equation by iteration. We have $g = 1$, and at the freeze-out temperature $g_* = 92$ (Kolb and Turner 1990). For masses close to the electroweak scale, x_F is of the order of 20 (see below for an estimate). Equation 2.7 therefore writes:

$$\Omega h^2 \simeq 0.1 \frac{3 \times 10^{-26} \text{cm}^3 \text{s}^{-1}}{\langle \sigma v \rangle}. \tag{2.10}$$

Following the measurements from Planck shown in the Table 2.1, this implies that the thermally averaged velocity weighted annihilation cross-section for DM $\langle \sigma v \rangle \sim 3 \times 10^{-26}$ cm^{3}s^{-1}. A careful calculation as performed in Steigman et al. (2012) gives $\langle \sigma v \rangle = 2 \times 10^{-26}$ cm^{3}s^{-1} with little dependence for electroweak-scale masses.

However, a simple estimate of x_F can be obtained. Given it is a log quantity, one can assume it as $\mathcal{O}(1)$. Therefore, with $g = 1$ and $g_* = 92$, Eq. (2.9) gives:

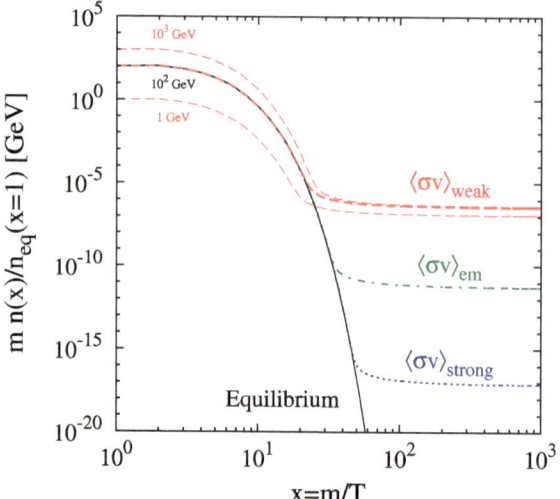

Fig. 2.5 Solution of the Boltzmann equation for masses of 1, 10^2 and 10^3 GeV, respectively. The solid black curve represents the equilibrium WIMP mass density $mn(x)/n_{eq}$. The curves show the WIMP mass density, normalized to the initial equilibrium number density, for different values of the annihilation cross section $\langle \sigma v \rangle$ and mass m. The density quickly reaches a constant after freeze-out which occurs at $x_F \simeq 20$ for electroweak scale cross section. Figure extracted from Steigman et al. (2012)

$$x_F \sim \ln(g\langle \sigma v \rangle m_{DM} M_{Pl}) - 3\log(2\pi) \sim 25, \tag{2.11}$$

for $m_{DM} = 1$ TeV and $\langle \sigma v \rangle = 3 \times 10^{-9}$ GeV^{-2}. With $\langle \sigma v \rangle \sim \alpha^2/m_{DM}^2$, a coupling typical to the electroweak scale $1/\alpha \simeq 30$, and $m_{DM} = 1$ TeV, naturally provide the correct thermal relic cross-section. This is commonly referred to as the WIMP miracle.

In the s-wave scenario, $\langle \sigma v \rangle$ is a constant in Eq. (2.5). More generally, $\langle \sigma v \rangle = \langle \sigma v \rangle_0 x^{-n}$ where $n = 0$ and 1 for s-wave and p-wave, respectively. For $x > x_F$, Y is much larger then Y_{eq} given that $Y_{eq} \propto \exp(-x)$. Therefore, given that $\Omega_{DM} \propto m_{DM} Y_\infty/\rho_c$, Eq. (2.5) gives:

$$\Omega_{DM} \propto \left(\int_{x_F}^{\infty} \langle \sigma v \rangle_0 \, x^{-(n+2)} \right)^{-1} = (n+1)x_F^{n+1}. \tag{2.12}$$

In order to keep Ω_{DM} fixed, one obtains that $\Omega_{DM} \propto (n+1)x_F^{n+1}$. Therefore, $\langle \sigma v \rangle_0^p = 2x_F\langle \sigma v \rangle_0^s = 1.2 \times 10^{-24}$cm^{3}s^{-1} using $x_F = 25$.

Figure 2.5 shows the solution of the Boltzmann equation for masses of 1, 10^2 and 10^3 GeV, respectively. The solid black curve represents the equilibrium WIMP mass density mn/n_{eq}. The curves show the WIMP mass density, normalized to the initial equilibrium number density, for different values of the annihilation cross section

$\langle \sigma v \rangle$ and mass m. The density quickly reaches a constant after freeze-out. The Figure was extracted from Steigman et al. (2012).

2.1.4 Lambda-CDM Issues at Small Scales and Alternatives

The existence of DM postulated by the ΛCDM is still a very successful solution to explain most of the cosmological and astrophysical measurements. The model agrees with observations for large scales, like at the level of galaxy clusters and in the primordial Universe. However, some issues not explained by the model arise for small-scale structures, *i.e.* at galactic scale. A pure CDM scenario incurs three main issues: (*i*) the missing Galactic satellites predicted from the model, (*ii*) the uncertainty on the cusp/core parameterization for the central galactic regions, (*iii*) the prediction for the galactic disks are too small compared to observations—known as the angular momentum problem. These problems may arise due to the unknown DM nature or from the simulations' lack of spatial resolution or realism when dealing with physical processes. It is however worth noticing that modifications to the collisionless cold DM paradigm may be unnecessary with more observational data and a better understanding of structure formation in the presence of baryons.

(*i*) The missing satellites problem

When quoting the *missing satellites problem*, one refers to the lack of observable satellite galaxies in the Local Group with respect to what is predicted by the CDM model (Klypin et al. 1999; Moore et al. 1999). These predicted subhalos are not observed and, therefore, can only be described as dark halos (Chiu et al. 2001). This is interpreted as if the smallest dark matter halos were extremely inefficient at forming stars. The discrepancy can also be alleviated with a part of the dark matter being warm or self-interacting.

The merging process is likely to spare substructures within CDM halos, as indicated by previous research (Kauffmann et al. 1993). Halos comparable in size to the Milky Way are expected to harbor numerous satellite subhalos, with the potential for more than ~ 100 objects reaching sufficient mass to support observable satellite galaxies (with $L > 10^6 L_\odot$). However, at that time, only approximately ~ 10 satellites brighter than this were observed around the Milky Way. This leads to the inference that the formation of dwarf galaxies would have to be significantly inhibited to account for this inconsistency. In the late 1990s, numerical simulations corroborated the issue of dwarf satellites (Klypin et al. 1999; Moore et al. 1999). Recent simulations have reaffirmed these findings (refer, for example, to Springel et al. (2008), Diemand et al. (2008)). The substructure's mass function is anticipated to sharply increase at the smallest masses, while the observed dwarf satellite luminosity function remains relatively flat.

Over the past decade, a pivotal observational development has been the identification of a novel group of faint satellite galaxies. This discovery has more than doubled the previously recognized satellite population within the Local Group, as outlined in

Belokurov et al. (2009). One intriguing aspect of these findings is the revelation that a substantially larger number of undiscovered dwarf galaxies exist as missing satellites within the Milky Way's halo. These elusive entities are predominantly composed of dark matter, exhibiting extremely faint and diffuse characteristics that have, until recently, eluded detection. It is conceivable that a considerable quantity, potentially exceeding one hundred, of these low-luminosity galaxies may be in orbit within the Milky Way's halo. The diffuse stellar distributions of these systems have, thus far, rendered them challenging to detect; however, ongoing surveys and advancements in techniques hold promise for unveiling hundreds of these previously missing satellite galaxies in the coming decade. Low-luminosity satellite galaxies have been identified by recent surveys like DES (Drlica-Wagner et al. 2015). Ultra-faint dwarf galaxies are expected to be found by new surveys like PanStarrs and SkyMapper.

Nevertheless, there exist physical mechanisms that could potentially inhibit the formation of galaxies within the smallest halos. Specifically, it is understood that feedback processes, notably the early reionization of gas caused by the first stars and the winds produced by supernovae, impede the formation of visible small galaxies from the vast majority of small subhalos that endure within large CDM halos (Somerville 2002; Bullock et al. 2000; Benson et al. 2003).

An alternative to the missing satellite problem from particle physics is warm dark matter (WDM). In the WDM scenario, the structures of comoving sizes smaller than l_0 are less abundant as compared to CDM. This is the region's size from which matter collapses into a compact object. Requiring that the mass of DM, initially distributed over the volume of comoving size l_0, which collapses later on, is of the order of the mass of a dwarf galaxy M gives

$$l_0 = \left(\frac{3M}{4\pi \Omega_{\mathrm{DM}} \rho_c} \right)^{1/3} \sim 100 \, \mathrm{kpc} \,, \tag{2.13}$$

with $M = 10^8 M_\odot$. Before becoming non-relativistic, WDM particles travel a distance of the order of the horizon size given by $l_H(t_{\mathrm{NR}}) \sim H^{-1}(T \sim m)$ and therefore WDM perturbations are suppressed at those scales. The horizon size at the time t_{NR} when $T \sim m$ is of order $l_H(t_{\mathrm{NR}}) = H^{-1}(T \sim m) = M_{\mathrm{Pl}}/m^2$. Given the expansion of the universe, $l_0 \sim l_H(t_{\mathrm{nr}}) a_0 / a_{t_{\mathrm{NR}}} \sim M_{\mathrm{Pl}}/(mT_0)$. With $T_0 = 2.7$ K, one gets m_{WDM} of a few keV.

(ii) The core/cuspy tension

The central density of dark matter halos predicted by cosmological N-body simulations are modeled as a steep power-law-like behavior (Flores and Primack 1994; Moore 1994; Diemand et al. 2005; Springel et al. 2008). However, a core-like structure is deducted from the observed galaxy rotation curves (de Blok 2010). This is the *core/cuspy tension*. A core profile is also observed in dwarf irregular galaxies[3] dominated by dark matter.

---

[3] Dwarf irregular galaxies appear small, faint, unstructured, and irregular in shape at optical wavelengths. These systems present some level of star formation despite being low surface brightness,

These galaxies are dominated by dark matter, while the baryon mass is dominated by gas. The surface densities of dark matter and HI are distributed with the same radial profiles, in these objects (Hoekstra et al. 2001). An example of this phenomenon is the dwarf irregular galaxy DDO154 (Carignan and Beaulieu 1989). One potential explanation for the absence of a dark matter cusp in observations is that dark matter may not be the dominant component in the centers of these galaxies. It's worth noting that this is already observed in more mature early-type galaxies, where the central dominance is attributed to stars. Another possibility is that the mass at the centers of these galaxies, which are dominated by HI, might predominantly consist of baryonic matter, specifically in the form of cold, condensed molecular gas.

Addressing this issue demands considerable effort, yet no solution has proven entirely satisfactory. The study referenced in Milosavljević and Merritt (2001) demonstrated that black hole binaries have the capacity to flatten cusps, but such binaries are not known to exist in dwarf galaxies. Central cusps exhibit resilience to stellar feedback, as indicated by the findings in Gnedin and Zhao (2002), even when the density is moderately reduced. Bars, as discussed in Weinberg and Katz (2002), have a limited impact and are not likely to be present in dwarf galaxies.

Determining the behavior of DM density profiles in the central regions of galaxies faces significant uncertainties, primarily because the density of visible baryonic matter is anticipated to surpass that of DM at small galactocentric radii. On the observational front, this implies that transitioning from gravitational measurements of the total mass density to constraints on the DM density necessitates meticulous modeling of the baryonic components (such as stars and gas), accompanied by sizable systematic uncertainties. Predictions based on simulations, which incorporate hydrodynamics and feedback physics alongside gravitational effects to estimate the expected DM abundance, are marked by considerable uncertainties owing to the influence of baryonic physics at the smallest scales. Additionally, at sufficiently small galactocentric distances, the resolution limit of simulations becomes pertinent, further contributing to uncertainties.

Besides the astrophysical solutions to the cusp-core problem with baryons (supernovae feedback, etc.), and also with interactions of dwarf satellite galaxies with the large host galaxy, a particle physics proposal suggests that DM is cold, but elastic scattering of DM particles smoothes out the cuspy mass distribution at centers of galaxies. Assuming that a DM mass density of $\rho_{DM} \sim 1$ GeV/cm^3 and requiring that the mean free path of a DM particle is $l \sim 1$ kpc to match possible core-like distribution in the most massive dwarf galaxy satellite of the Milky Way, $p = l\sigma n \sim 1$ gives $\sigma/m \sim 1$ barn/GeV. Interestingly, this provides constraints for light DM.

(*iii*) The angular momentum problem

The discrepancy between the relatively small size of galaxy disks in cosmological simulations and their observed counterparts created the *angular momentum problem*. In the current framework, both baryons and dark matter initially possess compara-

---

gas-rich, and metal-poor. They differ from dwarf spheroidal galaxies because they present star formation and detectable gas. They may have endured distinct episodes of star formation.

ble angular momentum. However, as galaxies form through hierarchical merging, baryons lose their angular momentum to CDM through dynamical friction. The challenge lies in the premature concentration of baryons and the resultant formation of galaxy disks due to merger events.

Alternatively, disks could be shaped by gas accretion from large-scale filaments, a process that might occur later in the galaxy's formation. One potential resolution involves enhancing the efficiency of feedback processes during star formation. Supernovae, for instance, could supply sufficient energy to maintain disks with angular momentum in line with observational requirements. This, in turn, influences the distribution of dark matter, leading to the realistic formation of cores. Disk formation occurs later, reducing dark matter concentration in the central region of galaxies. It's imperative for the feedback process to persist until the late stages to prevent gas collapse.

Another approach involves accreting mass from filaments, where gas channels in a non-spherical manner, retaining angular momentum. Observations of numerous barred galaxies support this mechanism. Without external cold gas accretion, bars can drive gas toward the galaxy center, experiencing substantial dynamical friction with early-condensed dark matter particles. A potential solution is a gradual and loosely-bound accretion of gas in the outer galaxy regions, mitigating the challenges associated with early and condensed gas collapse.

Alternatives to ΛCDM

Some theories have been developed to solve the difficulties that ΛCDM faces at the galactic scale. These theories exclude DM from the scenario explaining the Universe's formation, but can still explain some of the observational probes, like the dynamics of stars in the galaxies. Newton's law of gravity is modified in these theories (Milgrom 1983), and therefore, they are usually referred to as MOND (Modified Newtonian Dynamics) (Milgrom 1983). They are limited at galaxy clusters and cosmological scales even though they can well explain some effects at Galactic scales. Recent studies of gravitational waves and precise measurements of the speed of light have recently ruled out many of these theories (Boran et al. 2018).

2.2 Candidates to Explain Dark Matter

2.2.1 Primordial Black Holes

Primordial black holes (PBHs)—formed via the collapse of the large overdensities in the early Universe—have been studied since the '60 s (Zel'dovich and Novikov 1967) and are potential DM candidates (Hawking 1971; Chapline 1975). They could evaporate via the Hawking radiation (Hawking 1975, 1974), but the lifetime of PBHs with initial mass $M_{PBH} \gtrsim 10^{14}$ g is longer than the age of the Universe (Page 1976; MacGibbon et al. 2008). PBHs would behave as DM particles on cosmological scales;

however, on galactic or smaller scales, their granularity can produce observable effects. The 2-year MACHO collaboration's results on observations of microlensing on the Large Magellanic Cloud generated interest on PBHs in the late '90 s. More events than expected from known stellar populations were observed, and the found excess consisted of roughly half of the Milky Way halo. Astrophysical compact objects were excluded by arguments connected to the baryon budget (Fields et al. 2000). With later observations, the allowed halo fraction decreased (Alcock et al. 2000). Most of the theories and models for PBH DM date back to these years. After the LIGO-Virgo discovery of gravitational waves in 2016 from Solar mass black holes (Abbott et al. 2016), PBH DM reached a new level of interest thanks to the possibility that these BHs could be primordial rather than astrophysical (Bird et al. 2016; Clesse and García-Bellido 2017). A significant refinement for the abundance of PBHs came later. New constraints have been computed, with certain preexisting restrictions either lessened or eliminated. Theoretical computations of the processes behind the PBHs formation have undergone substantial enhancements. A thorough exploration of PBHs as potential candidates for DM is available in Carr et al. (2016). A detailed review on observational constraints over non-evaporated PBHs is available in Khlopov (2010).

Recent results on searches for bursts of γ-rays in TeV with timescales of a few seconds—expected from the PBHs evaporation in the final stage—have been obtained with observations with Cherenkov telescopes and are shown in Tavernier et al. (2020).

2.2.2 Axion, Axion-Like Particles, and Heavy Neutrinos

As a possible solution to the absence of CP violation in strong interaction, axions were introduced as particles. They can be predicted from QCD theories from non-zero quark masses. The solution to the strong CP problem from Peccei and Quinn introduces axions as bosons in the pseudo-Nambu-Goldstone theory (Peccei and Quinn 1977). In this solution, a U(1) approximate global symmetry is introduced, which is then broken at a scale f_a, located at around 10^{12} GeV. Axions couple to the standard matter as $\propto 1/f_a$. They are good candidates for DM because they are neutral, weakly interacting bosons. Even though they are very light, a population of non-relativistic—*i.e.*cold—axions can be produced when out of equilibrium (Duffy and van Bibber 2009). The approximate U(1) symmetry high-scale breaking can also generate axion-like particles (ALPs). These are not linked to the QCD theory; therefore, their mass and coupling to the standard matter are independent parameters and cannot be very well constrained with experiments. The interested reader can consult (Dark Sectors and New, Light, Weakly-Coupled Particles 2013) to review the search for axions and ALPs.

The standard left-handed neutrinos were postulated as DM candidates for hot DM up to a few eV in the late '70 s (Primack and Gross 2000). If the Universe were composed by hot DM, then a top-down formation scenario with superclusters formed first and fragmented later into smaller structures is necessary. However, this scenario does not reproduce the measured distribution of galaxies; hence, it is considered obsolete nowadays. On the other side, if a bottom-up formation scenario is assumed, neutrinos are wiped off before being able to form large-scale structures. Right-handed neutrinos are needed to account for the neutrino oscillations with a regular Dirac mass term added in the SM. Sterile neutrinos, interacting only via gravitational effects and not via weak interaction (this explains the name "sterile"), are hypothetical leptons (Mohapatra et al. 2007). In addition to the three left-handed SM active neutrinos interacting with W and Z bosons, right-handed sterile neutrinos not interacting with the electroweak bosons are present in four or more states. Hypothetically speaking, the masses of these states could be between 1 eV and 10^{15} GeV. The sterile neutrinos are tested for detecting neutrino oscillation anomalies at eV masses. They also serve as tests for baryogenesis theories at GeV-TeV masses. Sterile neutrinos at keV masses are good candidates for warm DM. This could also explain the formation of large-scale structures (Dolgov 2002). Merle (2013); Mohapatra et al. (2007) present all the previous cases.

2.2.3 Weakly Interacting Massive Particles

Important characteristics for a candidate DM particle are: non-baryonic, electromagnetically neutral, color neutral(-ish), massive (*i.e.* showing gravitational effects), living for a lifetime larger than the age of the Universe, reproducing the relic density measured with observations and sustaining the formation of the observed structures. A compelling candidate with these characteristics is WIMPs (as already introduced in Sect. 2.1.3). They interact gravitationally, and through any other forces—possibly not part of the Standard Model itself—with intensity either at the same level or lower than the weak nuclear force. They are also favored by a supersymmetric (SUSY) extension of the SM (Nilles 1984). WIMPs naturally reproduce the relic density of DM, a phenomenon known as the WIMP miracle—as discussed in Sect. 2.1.3.

According to SUSY, a supersymmetric partner with the difference of a half-integer spin exists for each particle: a supersymmetric fermion exists for each boson and vice versa. WIMPs candidates can be chosen among the superpartners of the bosons.

In SUSY, the protons decays with the process $p \rightarrow e^{+}\pi^{0}$; however, its timescale is rejected by observations. To solve this problem, a new discrete symmetry R-parity is defined by $R = (-1)^{2S+3B+L}$, with S the spin, B the baryon number, and L the lepton number of a particle. The symmetry is conserved and prevents the proton decay. SM particles have $R = 1$, while their superpartner particles in SUSY have $R = -1$. A consequence of the R-parity is that the lightest supersymmetric particle (LSP) is a good candidate for WIMPs because it is stable and cannot decay into SM particles with an opposite parity. The lightest neutralino—the lightest mixture between fermionic partners of the neutral Higgs boson and neutral electroweak gauge bosons—is a particularly good candidate. The superpartners of the Higgs boson and the W boson are the Higgsino and the Wino, respectively. The Bino is the superpartner for the gauge boson of the U(1) gauge field corresponding to weak hypercharge. Moreover, neutralinos are Majorana fermions; therefore, they can self-annihilate since each particle is identical to its antiparticle. They interact via the weak vector bosons. Heavy neutralinos can produce the lightest neutralino through the Z boson decay, which is then visible in a detector together with a missing momentum in the final state of the interaction.

Using the measured thermal relic DM density, the mass of the WIMP candidates can be constrained. The needed mass to reproduce the former and the thermal relic cross-section, is usually called the thermal mass. A pure Wino candidate would have a 2.9 ± 0.1 TeV thermal mass. A pure Higgsino 1.0 ± 0.1 TeV. The Quintuplet, another possible state for WIMP dark matter, is the 5 representation of SU(2) which would result in a thermal mass of 13.6 ± 0.8 TeV (Hisano et al. 2007; Bottaro et al. 2022; Mitridate et al. 2017; Cirelli et al. 2007).

This range of masses is within the sensitivity of indirect searches for DM with gamma-ray telescopes. Alternative candidates to SUSY are particles in the Kaluza-Klein theory (KK). These are theorized for a multidimensional Universe (Kaluza 1921), consisting of the brane 4-dimensional Universe embedded in a $3 + \delta + 1$-dimension space-time called bulk. The KK particles are the states propagating through the small extra dimensions and operate like partners of SM particles, but with the same spin as opposed to SUSY particles. A new discrete KK-parity is introduced similarly to the R-parity in SUSY. A good DM candidate, an alternative to the LSP, is the Lightest KK particle (LKP) (Kolb and Slansky 1984).

2.3 Some Bounds on the Particle Dark Matter Models

2.3.1 Bounds on Dark Matter Mass

The Fermi energy for an ensemble of non-interacting and non-relativistic fermions of mass m_f and density n_f is given by $E_f = (\hbar/2m_f)(3\pi^2 n_f)^{2/3}$. Assuming a constant-density sphere of mass M and radius R filled with identical fermions and $E_f = m_f v^2/2$, one gets $v_f = \hbar(9\pi M/4m_f R^3)^{1/3}$. To keep the sphere stable and avoid evaporating due to Fermi degeneracy pressure, one requires the $v_f \leq v_{esc} = \sqrt{2GM/R}$, which enables to derive a lower limit on the fermionic DM mass given by Boyarsky et al. (2009):

$$m_f^4 \geq \left(\frac{9\pi}{4}\right)\frac{\hbar^3}{2\sqrt{2}R^{3/2}G_N^{3/2}M^{1/2}}. \tag{2.14}$$

For a typical dwarf galaxy taken as a sphere of mass $M = 10^8 \, M_\odot$ and $R = 1$ kpc, one gets that the mass for fermionic DM is $m_f \gtrsim 0.1$ keV. A more careful derivation has been pioneered in Tremaine and Gunn (1979) and is known as the Tremaine-Gunn bound.

For DM masses lower than keVs, the DM needs to be a boson because the quantum occupation numbers will be larger than one in dense systems like dwarf galaxies. For bosonic DM, an estimate is based on the uncertainty principle. The DM velocity cannot be determined to greater precision than $v_b \gtrsim 1/(R\,M_b) \sim 20$ km s^{-1} (10^{-22} eV$/m_b$). For dwarf galaxies, the escape velocity is about 20 km s^{-1}. To get the formation of DM structures like dwarf galaxies, $m_b \gtrsim 10^{-22}$ eV. Interestingly, current constraints from Lyman-α forest lies at the level of 10^{-21} eV (Iršič et al. 2017). DM mass made of very light bosons is sometimes called fuzzy dark matter.

2.3.2 Bounds on Dark Matter Self-Interaction

The most popular example of dark matter interaction comes from the observation of the Bullet Cluster (see Sect. 2.1). The two DM halos passed through unaffected, which means that the DM did not undergo significant self-interactions ($DM\,DM \rightarrow DM\,DM$) during the crossing, unlike the gas.

To estimate σ, we assume that the two galaxy clusters have a mass $M = 5 \times 10^{14} M_\odot$ and size $L = 1$ Mpc. The probability of interaction can be estimated as $p \sim n\sigma L$, with the DM density given by $n = 3M/(4\pi L^3 m)$ and m the DM mass. In order to avoid that most of DM undergoes scattering, $p < 1$, which implies that $\sigma/m \simeq 2$ cm^2/g $\simeq 4$ barn/GeV. Note that this constraint is quite loose beyond GeV masses.

2.3.3 Bounds for the Weakly Interacting Massive Particles Mass

If one makes the assumption that resonance and coannihilation with other new particles slightly heavier than DM can be neglected, a standard thermal history for early universe cosmology suggests a natural value for the $\langle \sigma v \rangle$ of thermally produced DM particle at the order of $\sim 10^{-26}$ cm^{3}s^{-1}. This is independednt of the DM particle mass to logarithmic corrections.

The upper bound of the WIMP mass range is given by the unitarity limit, as derived in the seminal paper of Griest and Kamionkowski (1990) using partial-wave analysis. Unitarity for point-like self-conjugate Majorana fermion DM particle requires:

$$m_{\rm DM} \lesssim 100\,{\rm TeV} \left(\frac{3 \times 10^{-26}\,{\rm cm^3 s^{-1}}}{\sigma v_{\rm rel}} \right)^{1/2}. \tag{2.15}$$

A more refined estimates provides $m_{DM} < 194\,\text{TeV}$ (Smirnov and Beacom 2019; Tak et al. 2022). Of course, one can evade the unitarity bound—even without modifying the underlying cosmology—if the particle physics is modified to include compositeness or bound state formation. In this case, much heavier DM can be accommodated (see, e.g., Carney et al. (2023), Tak et al. (2022), Bottaro et al. (2022)).

Several DM candidates have been proposed in the literature Bergström (2009), Feng (2010).

The properties of DM candidates are mostly unconstrained. The total $\langle \sigma v \rangle$ is determined by the relic abundance, but the branching ratios to specific annihilation channels are model-dependent. When no preferred model is chosen, searches for a DM signal have to be conducted in a model-independent approach as far as possible. To satisfy this purpose, the constraints on DM parameters, that will be presented in the later chapters, will be shown for a wide range of DM particle masses and annihilation spectra.

If DM particles are thermally produced in the early Universe and comprise all the DM abundance, there are two important constraints on its properties that one can derive from theory. This comes from the unitarity of the scattering matrix (Griest and Kamionkowski 1990; Hui 2001). One obtains the upper bound on $\langle \sigma v \rangle$ by following the formalism of quantum field theory in the two-body scattering process and applying an expansion in partial waves. If one assumes a low-velocity limit, where s-wave annihilation dominates, this can be obtained:

$$\langle \sigma v \rangle = \frac{4\pi}{m_{DM}^2 v} \simeq 1.5 \times 10^{-19} \text{cm}^3\text{s}^{-1} \left(\frac{1\,\text{TeV}}{m_{DM}} \right)^2 \left(\frac{300\,\text{km s}^{-1}}{v} \right). \tag{2.16}$$

This is often referred to as the *unitarity bound*. Therefore, the observed CDM relic density places an upper bound on the DM mass.

The requirement that the DM annihilation does not significantly distort the DM halos in today's Universe places the second constraint. It is interesting to derive how large the annihilation can be irrespective of possible constraints from the early Universe, particularly the galaxies DM profiles. If one takes the cross section from the KKT model in Kaplinghat et al. (2000),[4] this can be interpreted as an upper bound corresponding to a value of $\langle \sigma v \rangle$ which produces significant distortion in the halo. This can be expressed as:

$$\langle \sigma v \rangle_{KKT} \lesssim 3 \times 10^{-16} \text{cm}^3\text{s}^{-1} \left(\frac{m_{DM}}{1\,\text{TeV}} \right). \tag{2.17}$$

This bound is significant for relatively small DM masses.

---

[4] In the context of the core-cusp problem, large annihilation rates were in fact invoked to alter the DM halos density profile, solving an apparent discrepancy between predicted sharp cuspy profiles and the observed flat cores. These similar effects are obtained with self-interacting DM.

2.4 Detection Techniques for WIMP Dark Matter

2.4.1 Production at Colliders

To explore DM at particle colliders, DM particles are generated via interactions involving SM particles accelerated in processes denoted as $XX \rightarrow \chi\chi$, considering X as a SM particle and χ as a DM particle. The Run 3 at the Large Hadron Collider (LHC), where protons (pp) collide at a center-of-mass energy of 13.6 TeV, offers the potential to accumulate sufficient data and luminosity to impose highly restrictive constraints on DM searches. No candidate DM particles have been directly detected yet at the moment of the writing, but these collider experiments have yielded stringent limits on certain DM models.

In collider experiments, the DM particles produced are not directly observed. Instead, the missing energy or missing transverse momentum serves as a smoking gun signature (Fox et al. 2012). In the context of theories beyond the Standard Model (BSM), scenarios often consider a single DM particle without additional BSM particles. Mediators, such as the Z boson or the Higgs boson, are postulated to facilitate interactions between SM and DM particles. However, when these mediators are much heavier than the collision energy, the interactions between DM and SM particles become contact-like, allowing for simplified models. These models are rooted in effective field theories (EFTs) (Liem et al. 2016), which reduce the assumptions about DM properties, such as its coupling with SM particles.

Alternatively, some simplified models are constructed by assuming that the mediator will primarily decay into SM partons, leading to descriptions of the visible physics in the final state, with less emphasis on higher-energy physics beyond the collider scale (Abdallah et al. 2015). More complex models may be applied for specific channels, incorporating additional information about distinct characteristics and signatures.

At the LHC, several benchmark channels are employed in the pursuit of DM: (*i*) Production through Z bosons with invisible decay, characterized by substantial missing transverse momentum and occasionally a single photon from initial state radiation (ISR). (*ii*) Production via the Higgs boson, followed by decay into a pair of Z bosons that subsequently undergo invisible decay. (*iii*) More general scenarios featuring heavy invisible particles decaying with the mediation of Z or Higgs bosons, resulting in signatures involving missing energy and ISR, such as mono-jet and mono-Higgs processes. (*iv*) Production of mediators alongside two top or bottom quarks, resulting in multi-jets in addition to missing energy. (*v*) More intricate, specific channels involving the production of Supersymmetric (SUSY) particles, often characterized by missing transverse momentum. (*vi*) Detection of vertices associated with displaced decay of long-lived particles (LLP) or complex signatures arising exclusively in the external sub-detectors, including calorimeters and muon spectrometers.

It's important to note that claiming a DM discovery based solely on collider data requires confirmation from direct or indirect search methods. Nonetheless,

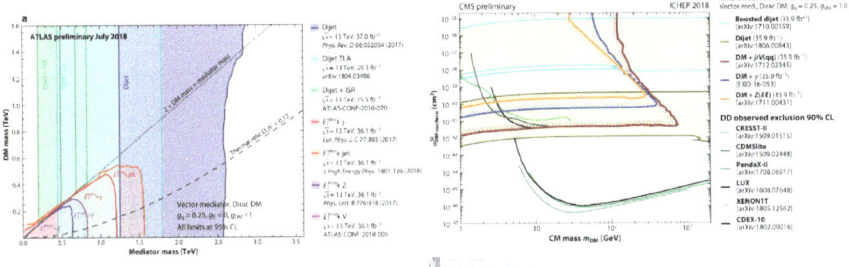

Fig. 2.6 Summary for the search of DM at colliders with specific models for ATLAS and CMS. *Left panel*: 95% C.L. ATLAS constraints in the DM mass vs mediator mass region. Combinations of masses consistent with the relic DM density measurements are given by the dashed line. *Right panel*: 95% C.L. CMS constraints on the DM-nucleus cross section, spin-independent and as a function of DM mass. Constraints from direct searches are also shown. Figures extracted from Boveia and Doglioni (2018)

these collider experiments offer the potential to discover new BSM particles. Figure 2.6 summarizes the constraints obtained through experiments conducted by ATLAS and CMS for specific DM models—at the moment of the writing, alongside a comparison with constraints from direct DM searches. For more comprehensive insights into DM searches at colliders, refer to Boveia and Doglioni (2018).

2.4.2 Direct Search

The direct search for DM consists in investigating phenomena where a DM particle χ interacts directly with a particle X from the SM in a process denoted as $\chi X \to \chi X$. Experiments investigating this measure the recoil of the nucleus of a target material due to elastic scattering with galactic WIMPs. The likelihood of observing a signal is contingent upon the DM particle's mass and the interaction cross-section between DM particles and the target material. Additional critical factors to consider include the local DM density and the velocity distribution of DM particles in the Milky Way, the latter being considerably uncertain.

The energy spectrum for the nuclear recoil can be expressed as $dR/dE_R \sim R_0/(E_0 r)e^{-E_R/E_0 r}$, as discussed in Lewin and Smith (1996). Here, E_R represents the recoil energy, E_0 represents the kinetic energy of the incoming DM particle, and R and R_0 denote the event rate per unit mass and the total event rate, respectively. The kinematic factor r depends on the masses of the target nucleus (m_T) and the DM particle (m_{DM}), defined as $r = 4m_T m_{DM}/(m_T + m_{DM})^2$. By measuring R at a specific E_R and fixing m_{DM}, one can constrain the DM signal rate, which in turn allows us to derive limits on the elastic-scattering cross-section of DM off nucleons.

To illustrate, assuming a Galactic velocity on the order of 10^{-3} times the speed of light and DM masses ranging from 10 GeV to 1 TeV, one can anticipate a recoil energy in the range ~ 1–100 keV. This corresponds to an expected differential rate at Earth of about 1 event per keV per kg per day (Lewin and Smith 1996). Nevertheless, direct DM detection faces significant challenges due to the rarity of recoil events. Moreover, there is a need to enhance sensitivity to low-mass DM by lowering the detection threshold.

Several techniques can be employed to discern a true DM signal from sources of background, such as analyzing pulse shapes. Background sources include electron recoils from external gamma-ray radiation, contamination signals within the detector, and elastic scattering of neutrinos from the Sun. Discriminating nuclear recoils from fission events poses a more challenging and sometimes insurmountable task. Additional background sources include recoils of alpha particles, interactions between atmospheric muons and neutrons, and coherent scattering of neutrinos with nuclei. Efforts to mitigate background interference involve shielding and placing detectors in underground laboratories and utilizing low-background materials.

The ideal detectors for direct DM detection should employ target nuclei with a large mass, a low threshold for recoil energy (E_R), minimal background interference, and the ability to distinguish between nuclear and electron recoils. Existing detectors utilize diverse materials, target nuclei, and detection techniques. For instance, noble liquid targets are utilized for large target sizes with low background, while cryogenic crystal targets offer a low E_R threshold and high energy resolution. Detection methods often involve scintillation, ionization, low-temperature photon techniques, or combinations thereof. Some experiments, such as Darkside and XENON, employ liquid argon and xenon as targets and utilize both ionization and scintillation techniques. Other experiments, like DAMA/LIBRA, employ scintillators with NaI(Tl), SuperCDMS and EDELWEISS utilize cryogenic germanium and silicon detectors, and CUORE employs bolometers with tellurium.

Some experiments, like DAMA/LIBRA, search for an annual modulation of the count rate, which arises from the variation in the distance between the center of the Milky Way and the detector due to the Earth's orbit around the Sun. This technique relies on the Earth's motion, with count rate peaks expected when the relative velocity reaches its maximum in June. Notably, DAMA observed a significant signal (Bernabei et al. 2018), although this observation faced stringent constraints from other experiments.

An overview of the current constraints—at the moment of the writing—on the spin-independent WIMP-nucleon cross-section derived from direct DM searches is shown in the left panel of Fig. 2.7. These constraints pose challenges to the coherent elastic neutrino-nucleus scattering cross-section, commonly referred to as the neutrino floor (Bœhm et al. 2019), which constitutes an irreducible background in direct DM detection. Nevertheless, certain models below this threshold can be explored through indirect DM detection methods. Additional details regarding DM searches via direct detection can be found in the comprehensive review provided in Zyla et al. (2020).

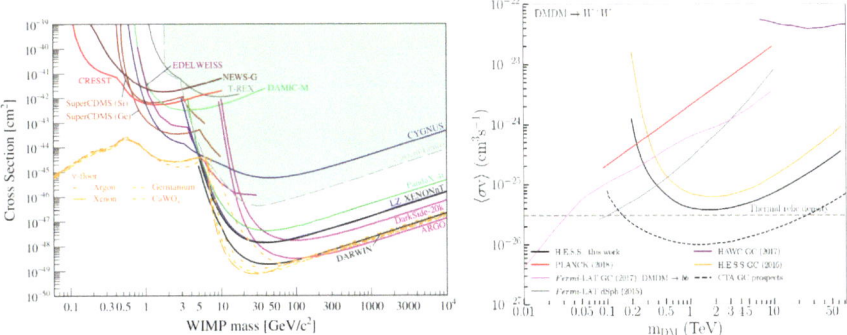

Fig. 2.7 *Left panel*: This summary focuses on the limits imposed by direct detection experiments on the spin-independent elastic cross-section between WIMPs and nucleons. The green-shaded area represents the portion of the parameter space presently ruled out by the collective sensitivity of all ongoing experiments. The colored lines on the chart indicate the anticipated sensitivities for either forthcoming experiments or upgrades to existing ones. Additionally, the orange-dashed line denotes the neutrino floor, which corresponds to the elastic cross-section of neutrino-nucleus interactions. Figure extracted from Billard et al. (2021). *Right panel*: Summary of constraints on WIMP self-annihilation cross section from indirect detection techniques. The limits indicated as the solid black line are also discussed in Chap. 7. The thermal relic cross-section is indicated by the gray line, which represents the natural annihilation scale for thermally produced WIMPs. Figure extracted from Abdalla et al. (2022) with the addition of the prospect limits with the Cherenkov Telescope Array (CTA) extracted from Acharyya et al. (2021)

2.4.3 Indirect Search

The process of detecting secondary SM particles arising from self-annihilating DM particles, described as $\chi\chi \rightarrow XX$, is commonly known as an indirect search. Here, X can be represented by a photon, neutrino, hadron, lepton, or an electroweak boson. Specific instruments have been developed to detect the various outcomes of these final states. When considering gamma-rays as the final state, a notable advantage is that their path is not bent by magnetic fields, allowing them to emanate directly from their source. So, the telescopes can be precisely pointed toward the densest regions of the Universe where DM is concentrated. However, it is essential to deal with a significant astrophysical backgrounds. Additionally, interactions with the Extra-galactic Background Light attenuate the gamma-ray spectrum, limiting the detection to gamma-rays up to a redshift of $z = 1$. Neutrinos, much like gamma-rays, exhibit minimal deviation from the direction of their source and undergo few interactions. Therefore, they can be used to probe up to far distances. They primarily interact weakly with matter, so indirect searches employing neutrino telescopes rely on large-scale underwater and under-ice experiments, such as ANTARES and IceCube. These experiments ensure that detected muons originate from cosmic neutrinos rather than background sources. Neutrinos can be produced promptly in DM annihilation or as secondary products from the decay of leptons (and antileptons) in the final state. Furthermore, DM annihilation can generate pairs of gauge bosons, which subsequently decay into leptons and ultimately into neutrinos. In cases where the neutral Z gauge boson is produced, direct decay into neutrinos can occur. The multiple scattering of solar nuclei and DM represents a clean channel for the indirect search of DM using neutrinos. Inside the Sun, the captured DM particles annihilate producing SM particles, which subsequently decay into neutrinos. These elusive neutrinos eventually escape the Sun and reach detectors on Earth (Adrian-M.ez et al. 2016; Aartsen et al. 2017). However, significant challenges are faced in neutrinos detection. Indirect searches for DM can also be conducted through satellite experiments such as AMS and PAMELA, which detect charged cosmic rays (CRs). In these experiments, the flux of electrons, protons, and their antiparticles is measured. At GeV energies, charged CRs are deflected by the Galactic magnetic field, resulting in an isotropic distribution that does not provide directional information unless measurements of very nearby sources are available. Therefore, indirect searches primarily focus on identifying an overall surplus of positrons and antiprotons compared to what can be explained by standard astrophysical processes. The search for antimatter benefits from a relatively low background. PAMELA, for instance, observed an excess of positrons (Adriani et al. 2009), a finding later confirmed by AMS-02 with enhanced precision and over a broader energy range (Aguilar et al. 2013). While these measurements may suggest a DM signal (Profumo and Jeltema 2009), they could also be explained by standard astrophysical processes associated with the acceleration of CRs in pulsars (Serpico 2012). To verify a DM hypothesis, further confirmation is needed from other experiments and measurements, particularly concerning the flux of antiprotons and gamma-rays. Notably, an excess of antiprotons has been observed in measurements conducted by AMS Cuoco et al. (2017), hinting at the possible existence of DM particles with masses ranging from 40 to 130 GeV and thermal annihilation cross-sections. However, these findings are subject to uncertainties related to the propagation of CRs through the ISM. A summary of constraints derived from indirect DM searches through the detection of secondary SM particles can be found in the right panel of Fig. 2.7. For more comprehensive information on indirect DM searches, refer to Conrad and Reimer (2017).

2.4.4 Complementarity of the Detection Techniques

The different DM detection techniques are increasingly complementary, enhancing their collective capability to detect or significantly constrain DM within the GeV mass range. At the TeV scale, colliders face limitations due to their limited center-of-mass energy, and direct detection is hampered because DM particles with higher masses tend to be less abundant. However, the neutrino floor is now within reach of direct detection constraints. Figure 2.8 schematically represents the possible DM detection channels for the coupling between Standard Model and DM particles through an unknown interaction, illustrating the three experimental approaches.

Indirect detection becomes the preferred avenue for probing the TeV mass regime. Nevertheless, indirect detection encounters challenges related to contamination from standard astrophysical emissions. While uncertainties in the local DM density persist, the DM density distribution in observed targets is also known with limited certainty. Fundamental properties of DM, such as its spin and couplings, can be drawn through the production of DM particles at colliders. These properties, especially the spin, are often inaccessible via indirect detection methods. However, if a candidate DM particle is discovered at a collider, confirmation from indirect and direct detection techniques would be essential to confirm whether the DM in the Universe is indeed composed of this new particle. To compare the three detection techniques, it is necessary to understand the underlying DM interaction. EFTs and simplified models are employed in a model-dependent manner for this purpose (Meyer 2024). When the interaction's center-of-mass energy is significantly smaller than the mediator's mass, EFTs are suitable, with the mediator mass being integrated out, leaving only the DM particle as the relevant degree of freedom. Simplified models are employed when the EFT framework is not applicable, such as in the case of collider experiments, and these models include the mediator's properties in the calculations. Simplified models use specific Feynman diagrams, incorporating assumptions about the mediator's nature and its couplings to DM and Standard Model particles. Constraints on the parameter space, including the mediator mass and DM mass, derived from collider experiments, can subsequently yield constraints on the DM annihilation cross-section or DM-nucleon scattering cross-section without additional assumptions (CMS Collaboration 2024).

In Fig. 2.9, results from CMS are presented, with the interpretation based on a simplified model involving a pair of Dirac fermionic DM particles coupled to a mediator in the final state. The mediator can be of different types: vector, axial-vector, scalar, or pseudoscalar. A comparison between direct detection and collider searches is shown in the left panel, assuming a scalar mediator. For masses below 10 GeV, collider constraints greatly surpass those from direct detection. The right panel shows the comparison between indirect detection and collider searches, assuming a pseudoscalar mediator. In this case, collider searches reach constraints up to 200 GeV, and indirect detection constraints become more relevant in the same mass range. Constraints on $\langle \sigma v \rangle$ can be translated into the EFT scale M_* using EFT theories (Meyer 2024). Four operator cases are considered: scalar ($\mathcal{O}_S$), pseudo-scalar ($\mathcal{O}_P$), vector ($\mathcal{O}_V$), or axial-vector ($\mathcal{O}_A$). The operators $\mathcal{O}_P$ and $\mathcal{O}_A$ are suppressed by the target nucleus's spin or the scattering momentum exchange, while $\mathcal{O}_P$ and $\mathcal{O}_S$ are suppressed via a Yukawa coupling adhering to the principle of minimal flavor violation.

Figure 2.10 shows the sensitivity of direct underground searches with noble liquids compared to the current IACT limit and the expected future sensitivity. The sensitivity of IACTs provides a complementary means of testing the space of Minimal Supersymmetric Standard Model with 9 parameters (p9MSSM) space inaccessible to direct detection experiments due to the neutrino floor.

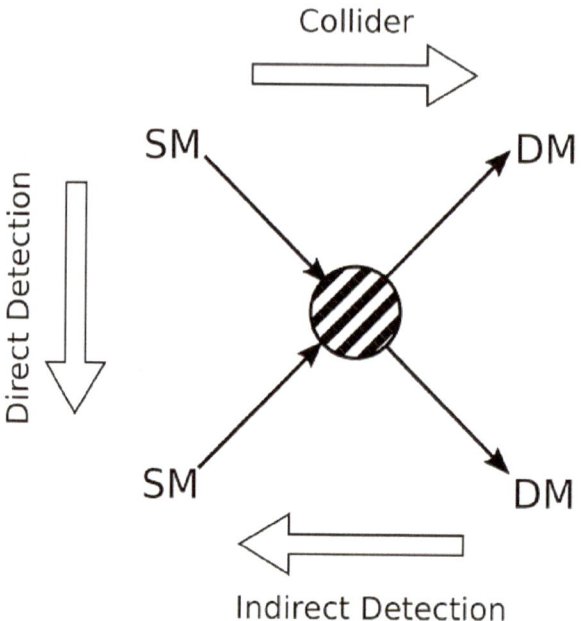

Fig. 2.8 The three experimental approaches for the possible DM detection channels represented schematically with the coupling of SM and DM particles through an unknown interaction, shown as the dash-shaded circle. Figure extracted from Marciano (2019)

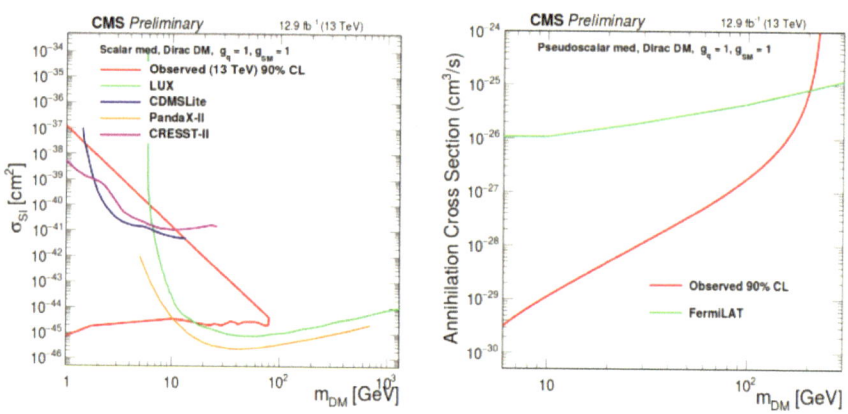

Fig. 2.9 Simplified models for the comparison of DM detection techniques. *Left panel*: comparison of direct detection and searches at collider assuming a scalar mediator. The constraints are for spin independent DM-nucleon scattering cross section versus DM mass. *Right panel*: comparison of indirect detection and searches at collider assuming pseudoscalar mediator. The constraints are for the DM annihilation cross section vs DM mass. Figure extracted from Meyer (2024)

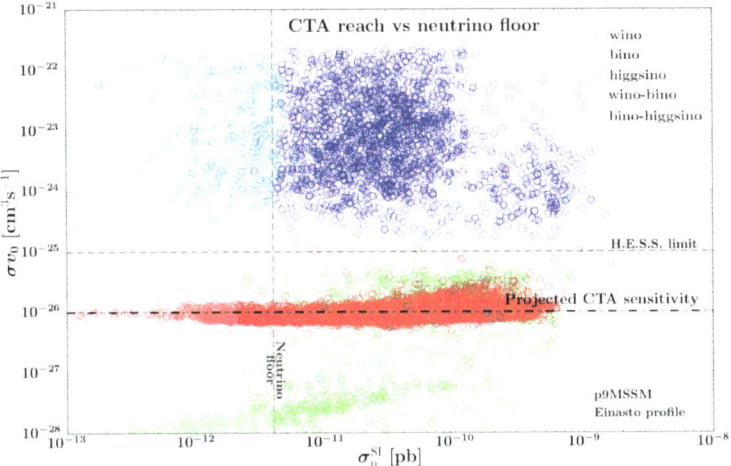

Fig. 2.10 p9MSSM points displayed in the $(\sigma_p^{SI}, \sigma_0 v)$ plane. The upper limit on σ_p^{SI} is given by the XENON1T sensitivity. The light-green and light-blue points are lying below the irreducible neutrino floor. The vertical line corresponds to the neutrino background limit for $m_\chi = 2$ TeV. The dashed horizontal line corresponds the 95% C.L. upper limit from H.E.S.S. taken at $m_\chi \simeq 2.5$ TeV, while the dashed double-dotted horizontal line shows the approximate CTA reach, taken at $m_\chi \simeq 1$ TeV. Figure extracted from Hryczuk et al. (2019)

References

Aartsen, M.G. et al.: Search for annihilating dark matter in the Sun with 3 years of IceCube data. Eur. Phys. J. C **77**.3 (2017). [Erratum: Eur.Phys.J.C 79, 214 (2019)], p. 146. https://doi.org/10.1140/epjc/s10052-017-4689-9

Abbott, B.P. et al.: Observation of gravitational waves from a binary black hole merger. Phys. Rev. Lett. **116**, 061102, 6 Feb. 2016. https://doi.org/10.1103/PhysRevLett.116.061102

Abdalla, H. et al.: Search for dark matter annihilation signals in the H.E.S.S. Inner galaxy survey. Phys. Rev. Lett. **129**.11, 111101 (2022). https://doi.org/10.1103/PhysRevLett.129.111101

Abdallah, J., et al.: Simplified models for dark matter searches at the LHC. Phys. Dark Univ. **9–10**, 8–23 (2015). https://doi.org/10.1016/j.dark.2015.08.001

Acharyya, A. et al.: Sensitivity of the Cherenkov Telescope Array to a dark matter signal from the Galactic centre. J. Cosmol. Astropart. Phys. **2021**.01, 057–057 (2021). https://doi.org/10.1088/1475-7516/2021/01/057. URL: https://doi.org/10.1088/1475-7516/2021/01/057

Adriani, O. et al.: An anomalous positron abundance in cosmic rays with energies 1.5–100 GeV. Nature **458**, 607–609 (2009). https://doi.org/10.1038/nature07942

Adrian-Mez, S. et al.: Limits on dark matter annihilation in the sun using the ANTARES Neutrino Telescope. Phys. Lett. B **759**, 69–74 (2016). https://doi.org/10.1016/j.physletb.2016.05.019

Aguilar, M. et al.: Composition of energy spectra and interactions, electrons, dark matter, neutrinos, muons, pions and other elementary particle detectors, cosmic ray detectors. Phys. Rev. Lett. **110**.14, 141102 (2013). https://doi.org/10.1103/PhysRevLett.110.141102

Alcock, C. et al.: The MACHO project: microlensing results from 5.7 years of large magellanic cloud observations. Astrophys. J. **542**.1, 281–307 (2000). https://doi.org/10.1086/309512

Baer, H. et al.: Dark matter production in the early Universe: beyond the thermal WIMP paradigm. Phys. Rept. **555**, 1–60 (2015). https://doi.org/10.1016/j.physrep.2014.10.002. arXiv: 1407.0017 [hep-ph]

Bassett, B.A., Hlozek, R.: Baryon Acoustic Oscillations, Oct. 2009. arXiv: 0910.5224 [astro-ph.CO]

Belokurov, V. et al.: The Discovery of Segue 2: a Prototype of the Population of Satellites of Satellites, vol. 397.4, pp. 1748–1755. https://doi.org/10.1111/j.1365-2966.2009.15106.x

Benson, A.J. et al.: The effects of photoionization on galaxy formation—III. Environmental Dependence in the Luminosity Function, vol. 343.2, pp. 679–691, Aug. 2003. https://doi.org/10.1046/j.1365-8711.2003.06709.x

Bergström, L.: Dark matter candidates. New J. Phys. **11.10**, 105006 (2009). https://doi.org/10.1088/1367-2630/11/10/105006.URL: https://dx.doi.org/10.1088/1367-2630/11/10/105006

Bergström, L.: Nonbaryonic dark matter: observational evidence and detection methods. Rept. Prog. Phys. **63**, 793 (2000). https://doi.org/10.1088/0034-4885/63/5/2r3

Bernabei, R. et al.: First model independent results from DAMA/LIBRA-phase2. In: Aharonov, Y. et al. (eds.), Universe, vol. 4.11, p. 116 (2018). https://doi.org/10.3390/universe4110116

Bertone, G., Hooper, D., Silk, J.: Particle dark matter: evidence, candidates and constraints. Phys. Rep. **405.5-6**, 279–390 (2005). https://doi.org/10.1016/j.physrep.2004.08.031

Billard, J., et al.: Direct Detect. Dark Matter—APPEC Committee Rep. (2021). https://doi.org/10.48550/ARXIV.2104.07634

Binney, J., Tremaine, S.: Galactic Dynamics, 2nd ed. (2008)

Bird, S. et al.: Did LIGO detect dark matter? Phys. Rev. Lett. **116**, 201301, 20 May 2016. https://doi.org/10.1103/PhysRevLett.116.201301

Bœhm, C. et al.: How high is the neutrino floor? J. Cosmol. Astropart. Phys. **2019.01**, 043–043 (2019). https://doi.org/10.1088/1475-7516/2019/01/043

Boran, S. et al.: GW170817 falsifies dark matter emulators. Phys. Rev. D **97.4**, 041501 (2018). https://doi.org/10.1103/PhysRevD.97.041501

Bottaro, S. et al.: Closing the window on WIMP Dark Matter. Eur. Phys. J. C **82.1**, 31 (2022). https://doi.org/10.1140/epjc/s10052-021-09917-9

Boveia, A., Doglioni, C.: Dark matter searches at colliders. Ann. Rev. Nucl. Part. Sci. **68**, 429–459 (2018). https://doi.org/10.1146/annurev-nucl-101917-021008

Boyarsky, A., Ruchayskiy, O., Iakubovskyi, D.: A lower bound on the mass of dark matter particles. JCAP **03**, 005 (2009). https://doi.org/10.1088/1475-7516/2009/03/005. arXiv: 0808.3902 [hep-ph]

Bullock, J.S., Kravtsov, A.V., Weinberg, D.H.: Reionization and the Abundance of Galactic Satellites, vol. 539.2, pp. 517–521, Aug. 2000. https://doi.org/10.1086/309279

Carignan, C., Beaulieu, S.: Optical and H i Studies of the "Gas-rich" Dwarf Irregular Galaxy DDO 154", vol. 347, p. 760, Dec. 1989. https://doi.org/10.1086/168167

Carney, D. et al.: Snowmass 2021 cosmic frontier white paper: ultraheavy particle dark matter. SciPost Phys. Core **6.4** (2023). ISSN: 2666–9366. https://doi.org/10.21468/scipostphyscore.6.4.075.URL: http://dx.doi.org/10.21468/SciPostPhysCore.6.4.075

Carr, B., Kuhnel, F., Sandstad, M.: Primordial black holes as dark matter. Phys. Rev. D **94.8**, 083504 (2016). https://doi.org/10.1103/PhysRevD.94.083504

Chapline, G.F.: Cosmological effects of primordial black holes. Nature **253.5489**, 251–252 (1975). https://doi.org/10.1038/253251a0

Chiu, W.A., Nickolay, Y. G., Ostriker, J.P.: The expected mass function for low-mass galaxies in a cold dark matter cosmology: is there a problem? Astrophys. J. **563.1**, 21 (2001). https://doi.org/10.1086/323685.URL: https://dx.doi.org/10.1086/323685

Cirelli, M., Strumia, A., Tamburini, M.: Cosmology and astrophysics of minimal dark matter. Nucl. Phys. B **787**, 152–175 (2007). https://doi.org/10.1016/j.nuclphysb.2007.07.023. arXiv: 0706.4071 [hep-ph]

Clesse, S., García-Bellido, J.: The clustering of massive Primordial Black Holes as Dark Matter: measuring their mass distribution with advanced LIGO. Phys. Dark Univ. **15**, 142–147 (2017). https://doi.org/10.1016/j.dark.2016.10.002

Clowe, D. et al.: A direct empirical proof of the existence of dark matter. Astrophys. J. **648.2**, L109–L113 (2006). https://doi.org/10.1086/508162

CMS Collaboration: Search for dark matter in final states with an energetic jet, or a hadronically decaying W or Z boson using 12.9 fb^{-1} of data at $\sqrt{s} = 13 TeV$. https://cds.cern.ch/record/2205746/files/EXO-16-037-pas.pdf

Coc, A., Vangioni, E.: Primordial nucleosynthesis. Int. J. Mod. Phys. E **26.08**, 1741002 (2017). https://doi.org/10.1142/s0218301317410026.URL: https://doi.org/10.1142%5C%2Fs0218301317410026

Conrad, J., Reimer, O.: Indirect dark matter searches in gamma and cosmic rays. Nat. Phys. **13.3**, 224–231 (2017). https://doi.org/10.1038/nphys4049

Croton (2013) Damn You, Little h! (Or, Real-World Applications of the Hubble Constant Using Observed and Simulated Data). Publ. Astron. Soc. Australia **30**, e052 (2013). https://doi.org/10.1017/pasa.2013.31

Cuoco, A., Krämer, M., Korsmeier, M.: Novel dark matter constraints from antiprotons in light of AMS-02. Phys. Rev. Lett. **118.19**, 191102 (2017). https://doi.org/10.1103/PhysRevLett.118.191102

Dark Sectors and New, Light, Weakly-Coupled Particles (2013). https://doi.org/10.48550/ARXIV.1311.0029.URL: https://arxiv.org/abs/1311.0029

de Blok, W.J.G.: The core-cusp problem. Adv. Astron. **789293**, 789293 (2010). https://doi.org/10.1155/2010/789293

Deruelle, N., Uzan, J.-P.: The Lambda-CDM model of the hot Big Bang. In: Relativity in Modern Physics. Oxford University Press, Aug. 2018. https://doi.org/10.1093/oso/9780198786399.003.0059

Diemand, J. et al.: Cusps in cold dark matter haloes. Mon. Not. R. Astron. Soc. **364.2**, 665–673 (2005). ISSN: 0035-8711. https://doi.org/10.1111/j.1365-2966.2005.09601.x.eprint: https://academic.oup.com/mnras/article-pdf/364/2/665/18663249/364-2-665.pdf. URL: https://doi.org/10.1111/j.1365-2966.2005.09601.x

Diemand, J., et al.: Clumps and streams in the local dark matter distribution. Nature **454**, 735–738 (2008). https://doi.org/10.1038/nature07153

Dolgov, A.: Massive sterile neutrinos as warm dark matter. Astropart. Phys. **16.3**, 339–344 (2002). https://doi.org/10.1016/s0927-6505(01)00115-3

Drlica-Wagner, A. et al.: Eight Ultra-faint Galaxy Candidates Discovered in Year Two of the Dark Energy Survey, vol. 813.2, p. 109 (2015). https://doi.org/10.1088/0004-637X/813/2/109

Duffy, L.D., van Bibber, K.: Axions as dark matter particles. New J. Phys. **11.10**, 105008 (2009). https://doi.org/10.1088/1367-2630/11/10/105008

Einstein, A.: The foundation of the general theory of relativity. Ann. Phys. **49.7** (1916). Hsu, J.-P., Fine, D. (eds.), pp. 769–822. doi:https://doi.org/10.1002/andp.19163540702

Feng, J.L.: Dark matter candidates from particle physics and methods of detection. Ann. Rev. Astron. Astrophys. **48.1**, 495–545 (2010). https://doi.org/10.1146/annurev-astro-082708-101659.eprint: https://doi.org/10.1146/annurev-astro-082708-101659. URL: https://doi.org/10.1146/annurev-astro-082708-101659

Fields, B.D., Freese, K., Graff, D.S.: Chemical abundance constraints on white dwarfs as halo dark matter. Astrophys. J. **534.1**, 265–276 (2000). https://doi.org/10.1086/308727

Flores, R.A., Primack, J.R.: Observational and Theoretical Constraints on Singular Dark Matter Halos, vol. 427, p. L1, May 1994. https://doi.org/10.1086/187350

Formation of the large-scale structure in the Universe: filaments. http://cosmicweb.uchicago.edu/filaments.html. Accessed 30 Sept. 2010

Fox, P.J. et al.: Missing energy signatures of dark matter at the LHC. Phys. Rev. D **85.5** (2012). https://doi.org/10.1103/physrevd.85.056011

Friedmann, A.: Über die Krümmung des Raumes. Zeitschrift fur Physik **10**, 377–386 (1922). https://doi.org/10.1007/BF01332580. Jan

Gelmini, G., Gondolo, P.: DM Production Mechanisms, pp. 121–141 Sept. 2010. arXiv: 1009.3690 [astro-ph.CO]

Gnedin, O.Y., Zhao, H.S.: Maximum feedback and dark matter profiles of dwarf galaxies. In: Monthly Not. R. Astronom. Soc. **333.2**, 299–306 (2002). ISSN: 0035-8711. https://doi.org/

10.1046/j.1365-8711.2002.05361.x.eprint: https://academic.oup.com/mnras/article-pdf/333/2/299/18411901/333-2-299.pdf. URL: https://doi.org/10.1046/j.1365-8711.2002.05361.x

Griest, K., Kamionkowski, M.: Unitarity limits on the mass and radius of dark-matter particles. Phys. Rev. Lett. **64**, 615–618 (1990). https://doi.org/10.1103/PhysRevLett.64.615URL: https://link.aps.org/doi/10.1103/PhysRevLett.64.615

Hawking, S.W.: Black hole explosions? Nature **248.5443**, 30–31 (1974). https://doi.org/10.1038/248030a0

Hawking, S.: Gravitationally collapsed objects of very low mass. Mont. Not. R. Astr. Soc. **152**, 75 (1971). https://doi.org/10.1093/mnras/152.1.75

Hawking, S.W.: Particle creation by black holes. Commun. Math. Phys. **43** (1975). Gibbons, G.W., Hawking, S.W. (eds.), [Erratum: Commun. Math. Phys. **46**, 206 (1976)], pp. 199–220. https://doi.org/10.1007/BF02345020

Hisano, J. et al.: Non-perturbative effect on thermal relic abundance of dark matter. Phys. Lett. B **646.1**, 34–38 (2007). https://doi.org/10.1016/j.physletb.2007.01.012

Hoekstra, H., van Albada, T.S., Sancisi, R.: On the apparent coupling of neutral hydrogen and dark matter in spiral galaxies. Monthly Not. R. Astronom. Soc. **323.2**, 453–459 (2001). ISSN: 0035–8711. https://doi.org/10.1046/j.1365-8711.2001.04214.x.eprint: https://academic.oup.com/mnras/article-pdf/323/2/453/4076900/323-2-453.pdf. URL: https://doi.org/10.1046/j.1365-8711.2001.04214.x

Hryczuk, A. et al.: Testing dark matter with Cherenkov light—prospects of H.E.S.S. and CTA for exploring minimal supersymmetry. JHEP **10**, 043 (2019). https://doi.org/10.1007/JHEP10(2019)043

Hui, L.: Unitarity Bounds and the Cuspy Halo problem. Phys. Rev. Lett. **86**, 3467–3470 (2001). https://doi.org/10.1103/PhysRevLett.86.3467. URL: https://link.aps.org/doi/10.1103/PhysRevLett.86.3467

Iršič, V. et al.: First constraints on fuzzy dark matter from Lyman-α forest data and hydrodynamical simulations. Phys. Rev. Lett. **119.3**, 031302 (2017). https://doi.org/10.1103/PhysRevLett.119.031302, arXiv: 1703.04683 [astro-ph.CO]

Kaluza, T.: Zum Unitätsproblem der Physik. Sitzungsber. Preuss. Akad. Wiss. Berlin (Math. Phys.) **1921**, 966–972 (1921). https://doi.org/10.1142/S0218271818700017

Kaplinghat, M., Knox, L., Turner, M.S.: Annihilating cold dark matter. Phys. Rev. Lett. **85**, 3335–3338 (2000). https://doi.org/10.1103/PhysRevLett.85.3335. URL: https://link.aps.org/doi/10.1103/PhysRevLett.85.3335

Kauffmann, G., White, S.D.M., Guiderdoni, B.: The Formation and Evolution of Galaxies within Merging Dark Matter Haloes, vol. 264, pp. 201–218, Sept. 1993. https://doi.org/10.1093/mnras/264.1.201

Kermack, W.O., McCrea, W.H.: On Milne's theory of world structure. Mont. Not. Royal Astr. Soc. **93**, 519–529 (1933). https://doi.org/10.1093/mnras/93.7.519. May

Khlopov, MYu.: Primordial black holes. Res. Astron. Astrophys. **10**, 495–528 (2010). https://doi.org/10.1088/1674-4527/10/6/001

Klypin, A. et al.: Where are the missing Galactic Satellites? Astrophys. J. **522.1**, 82 (1999). https://doi.org/10.1086/307643.URL: https://dx.doi.org/10.1086/307643

Kolb, E.W., Michael, S.T.: The Early Universe, vol. 69 (1990). ISBN: 978-0-201-62674-2. https://doi.org/10.1201/9780429492860

Kolb, E.W., Slansky, R.: Dimensional reduction in the early universe: where have the massive particles gone? Phys. Lett. B **135**, 378 (1984). https://doi.org/10.1016/0370-2693(84)90298-3

Lemaitre, G.: Un Univers homogène de masse constante et de rayon croissant rendant compte de la vitesse radiale des nébuleuses extra-galactiques. Annales de la Soci é t é Scientifique de Bruxelles **47**, 49–59 (1927). Jan

Lewin, J.D., Smith, P.F.: Review of mathematics, numerical factors, and corrections for dark matter experiments based on elastic nuclear recoil. Astropart. Phys. **6**, 87–112 (1996). https://doi.org/10.1016/S0927-6505(96)00047-3

Liem, S. et al.: Effective field theory of dark matter: a global analysis. JHEP **09**, 077 (2016). https://doi.org/10.1007/JHEP09(2016)077

MacGibbon, J.H., Carr, B.J., Page, D.N.: Do evaporating black holes form photospheres? Phys. Rev. D **78**, 064043 (2008). https://doi.org/10.1103/PhysRevD.78.064043

Marciano, A.: WIMP Dark Matter Searches With the ATLAS Detector at the LHC (2019). https://www.frontiersin.org/articles/10.3389/fphy.2019.00075/full

Markevitch, M.: Chandra observation of the most interesting cluster in the Universe (2005). https://doi.org/10.48550/ARXIV.ASTRO-PH/0511345

Markevitch, M. et al.: Direct constraints on the dark matter self-interaction cross section from the merging galaxy cluster 1E 0657-56. Astrophys. J. **606.2**, 819–824 (2004). https://doi.org/10.1086/383178

Massey, R., Kitching, T., Richard, J.: The dark matter of gravitational lensing. Rept. Prog. Phys. **73**, 086901 (2010). https://doi.org/10.1088/0034-4885/73/8/086901

Merle, A.: keV neutrino model building. Int. J. Mod. Phys. D **22.10**, 1330020 (2013). https://doi.org/10.1142/s0218271813300206

Meyer, M. et al.: Future constraints of dark matter effective field theories and simplified models with the Cherenkov Telescope Array. https://indico.cern.ch/event/623880/contributions/2523949/attachments/1438677/2213569/mmeyer_cta_eft.pdf

Milgrom, M.: A modification of the Newtonian dynamics as a possible alternative to the hidden mass hypothesis. Astrophys. J. **270**, 365–370 (1983). https://doi.org/10.1086/161130. July

Milosavljević, M., Merritt, D.: Formation of galactic nuclei. Astrophys. J. **563.1**, 34 (2001). https://doi.org/10.1086/323830.URL: https://dx.doi.org/10.1086/323830

Mitridate, A. et al.: Cosmological implications of dark matter bound states. JCAP **1705.05**, 006 (2017). https://doi.org/10.1088/1475-7516/2017/05/006. arXiv: 1702.01141 [hep-ph]

Mohapatra, R.N. et al.: Theory of neutrinos: a white paper. Rep. Prog. Phys. **70.11**, 1757–1867 (2007). https://doi.org/10.1088/0034-4885/70/11/r02.URL: https://doi.org/10.1088%2F0034-4885%2F70%2F11%2Fr02

Moore, B.: Evidence Against Dissipation-Less Dark Matter from Observations of Galaxy Haloes, vol. 370.6491, pp. 629–631, Aug. 1994. https://doi.org/10.1038/370629a0

Moore, B. et al.: Dark matter substructure within galactic Halos. Astrophys. J. **524.1**, L19 (1999). https://doi.org/10.1086/312287. URL: https://dx.doi.org/10.1086/312287

New Hubble image of galaxy cluster Abell 1689. https://esahubble.org/news/heic1317/. Accessed 30 Sept. 2010

Nilles, H.P.: Supersymmetry, supergravity and particle physics. Phys. Rept. **110**, 1–162 (1984). https://doi.org/10.1016/0370-1573(84)90008-5

Page, D.N.: Particle emission rates from a black hole: massless particles from an uncharged, nonrotating hole. Phys. Rev. D **13**, 198–206, 2 Jan. 1976. https://doi.org/10.1103/PhysRevD.13.198

Peccei, R.D., Quinn, H.R.: CP conservation in the presence of instantons. Phys. Rev. Lett. **38**, 1440–1443 (1977). https://doi.org/10.1103/PhysRevLett.38.1440

Peebles, P.J.E.: The Large-Scale Structure of the Universe (1980)

Penzias, A.A., Wilson, R.W.: A measurement of excess antenna temperature at 4080 Mc/s. Astrophys. J. **142**, 419–421 (1965). https://doi.org/10.1086/148307. July

Planck 2018 results. VI. Cosmological parameters. Astron. Astrophys. **641**, A6 (2020). https://doi.org/10.1051/0004-6361/201833910

Primack, J.R., Gross, M.A.K.: Hot dark matter in cosmology. In: Caldwell, D.O. (ed.), pp. 287–308, July 2000. arXiv: astro-ph/0007165

Primack, J.R.: Dark Matter Struct. Form. Universe (1997). https://doi.org/10.48550/ARXIV.ASTRO-PH/9707285

Profumo, S., Jeltema, T.E.: Extragalactic inverse compton light from dark matter annihilation and the pamela positron excess. JCAP **07**, 020 (2009). https://doi.org/10.1088/1475-7516/2009/07/020

Robertson, H.P.: Kinematics and World-Structure. 3. Astrophys. J. **83**, 257–271 (1936). https://doi. org/10.1086/143726

Rubin, V.C., Ford, W.K. Jr.: Rotation of the Andromeda Nebula from a spectroscopic survey of emission regions. Astrophys. J. **159**, 379 (1970). https://doi.org/10.1086/150317

Serpico, P.D.: Astrophysical models for the origin of the positron "excess." Astropart. Phys. **39–40**, 2–11 (2012). https://doi.org/10.1016/j.astropartphys.2011.08.007

Smirnov, J., Beacom, J.F.: Tev-scale thermal WIMPs: unitarity and its consequences. Phys. Rev. D **100**, 043029 (2019). https://doi.org/10.1103/PhysRevD.100.043029. URL: https://link.aps.org/ doi/10.1103/PhysRevD.100.043029

Smoot, G.F. et al.: Structure in the COBE differential microwave radiometer first-year maps. Astrophys. J. Lett. **396**, L1 (1992). https://doi.org/10.1086/186504

Somerville, R.S.: Can photoionization squelching resolve the substructure crisis? Astrophys. J. **572.1**, L23 (2002). https://doi.org/10.1086/341444.URL: https://dx.doi.org/10.1086/341444

Springel, V., Frenk, C.S., White, S.D.M.: The large-scale structure of the Universe. Nature **440**, 1137 (2006). https://doi.org/10.1038/nature04805. arXiv:astro-ph/0604561

Springel, V. et al.: Simulations of the formation, evolution and clustering of galaxies and quasars. Nature **435.7042**, 629–636 (2005). https://doi.org/10.1038/nature03597. arXiv: astro-ph/0504097 [astro-ph]

Springel, V. et al.: The Aquarius Project: the Subhaloes of Galactic Haloes, vol. 391.4 pp. 1685–1711 (2008). https://doi.org/10.1111/j.1365-2966.2008.14066.x

Springel, V. et al.: The aquarius project: the subhaloes of galactic haloes. Mon. Not. R. Astron. Soc. **391.4**, 1685–1711 (2008). ISSN: 0035-8711. doi:https://doi.org/10.1111/j.1365-2966.2008. 14066.x.eprint: https://academic.oup.com/mnras/article-pdf/391/4/1685/4881147/mnras0391- 1685.pdf. URL: https://doi.org/10.1111/j.1365-2966.2008.14066.x

Steigman, G., Dasgupta, B., Beacom, J.F.: Precise relic WIMP abundance and its impact on searches for dark matter annihilation. Phys. Rev. D **86**, 023506 (2012). https://doi.org/10.1103/PhysRevD. 86.023506

Tak, D. et al.: Current and future γ-ray searches for dark matter annihilation beyond the unitarity limit. Astrophys. J. Lett. **938.1**, L4 (2022). ISSN: 2041-8213. https://doi.org/10.3847/2041-8213/ ac9387.URL: http://dx.doi.org/10.3847/2041-8213/ac9387

Tavernier, T., Glicenstein, J.-F., Brun, F.: Search for primordial black hole evaporations with H.E.S.S. PoS ICRC2019, 804 (2020). https://doi.org/10.22323/1.358.0804

The 2dF Galaxy Redshift Survey: spectra and redshifts. In: Monthly Notices of the Royal Astronomical Society, vol. 328.4, pp. 1039–1063, Dec. 2001. https://doi.org/10.1046/j.1365-8711.2001. 04902.x

The Baryonic Density. https://ned.ipac.caltech.edu/level5/Sept02/Roos/Roos4.html. Accessed 30 Spt. 2010

The Sloan Digital Sky Survey: Technical Summary. Astron. J. **120.3**, 1579–1587 (2000). https:// doi.org/10.1086/301513. arXiv:astro-ph/0006396 [astro-ph]

Tremaine S., Gunn, J.E.: Dynamical role of light neutral leptons in cosmology. Phys. Rev. Lett. **42** (1979). Srednicki, M.A. (ed.), pp. 407–410. https://doi.org/10.1103/PhysRevLett.42.407

van Albada, T.S., et al.: Distribution of dark matter in the spiral galaxy NGC 3198. Astrophys. J. **295**, 305–313 (1985). https://doi.org/10.1086/163375. Aug

Weinberg, M.D., Katz, N.: Bar-driven dark halo evolution: a resolution of the cusp-core controversy. Astrophys. J. **580.2**, 627 (2002). https://doi.org/10.1086/343847.URL: https://dx.doi.org/10.1086/ 343847

Zel'dovich, Y.B., Novikov, I.D.: The Hypothesis of Cores Retarded during Expansion and the Hot Cosmological Model, vol. 10, p. 602 Feb. 1967

Zwicky, F.: On the masses of nebulae and of clusters of nebulae. Astrophys. J. **86**, 217 (1937). https://doi.org/10.1086/143864

Zyla, P.A. et al.: Review of particle physics. PTEP **2020.8**, 083C01 (2020). https://doi.org/10.1093/ ptep/ptaa104

Chapter 3
Framework for Indirect Dark Matter Search with Gamma Rays

Abstract This chapter describes the framework and its ingredients which are needed to search for a Dark Matter (DM) signal with gamma-rays. After a quick overview of how Very-High-Energy gamma-rays can be used to probe physics beyond the standard model, the expected gamma-ray flux from annihilating DM as well as the most promising astrophysical environments to seek it are described. A focus on the expected DM distribution at the Galactic scale is maintained since the analyses shown in the following Chapters make use of observations of the inner halo of the Milky Way. The derivation of the "J-factor"—the DM distribution in the observed target—with illustrative examples is presented, as well as the expected annihilation spectra from observations of the Milky Way for several DM annihilation channels. Some hints on the particle physics and astrophysical effects that can boost the DM signal are discussed.

Keywords Very-high-energy gamma-ray astrophysics · Indirect dark matter search · Milky Way dark matter distribution · J-factor · Dark matter annihilation flux

3.1 Very-High-Energy Gamma Rays as Messengers

Gamma-rays are a direct outcome of radiative processes involving very-high-energy Cosmic Rays (CRs). Therefore, investigating sources capable of accelerating CRs to energies ranging from TeV to PeV can offer valuable insights into the mechanisms responsible for particle acceleration. It also sheds light on the objects responsible for accelerating CRs to the energy levels associated with the cosmic ray "knee," ultimately, it provides clues about the very origin of Galactic cosmic rays. These sources often have connections to astronomical phenomena such as supernova remnants (SNR), pulsars (PSR), active galactic nuclei (AGNs), and supermassive black holes. In this context, both the H.E.S.S. collaboration (Abramowski et al. 2016) and the LHAASO experiment (Cao et al. 2021) have made significant contributions by reporting measurements of gamma-rays at energies reaching hundreds of TeV.

Moreover, signals from DM annihilation or decay can be searched for by observing very-high-energy (VHE, $E \gtrsim 100$ GeV) gamma rays. This section briefly reviews what kind of physics analysis can be performed using VHE gamma-ray data.

3.1.1 Extragalactic Background Light

The Universe is non-transparent to photons because they traverse through the medium, engaging with background light and giving rise to electron-positron pairs. Consequently, the absorption of photons stemming from interactions with background radiation becomes an area of investigation. Within the energy range of GeV to TeV, gamma-rays can encounter absorption by the **Extragalactic Background Light** (EBL), while ultra-high-energy gamma-rays (> 100 TeV) may also interact with the Cosmic Microwave Background (CMB). The degree of attenuation in the gamma-ray spectrum is quantified by a factor represented as $\exp(-\tau(E, z))$, depending on the gamma rays' energy and the source distance z by the optical depth $\tau(E, z)$. Standard EBL models, such as the one from Franceschini and Rodighiero (2017), state that gamma-rays with an energy of 10 TeV exhibit an optical depth of approximately 0.5 when originating from sources situated at a redshift of $z = 0.01$ (equivalent to a distance of approximately 45 Mpc). This optical depth increases significantly, reaching approximately 100, for sources located as far away as $z = 1$ (approximately 3 Gpc).

3.1.2 Lorent Invariance Violation

The speed of light can exhibit variations with respect to energy, primarily due to modifications in the photon dispersion relation as hypothesized by certain quantum gravity models. Researchers investigate this phenomenon, known as **Lorentz Invariance Violation** (LIV), by observing VHE transient and short-lived events such as gamma-ray bursts (GRBs), flares from AGNs, or PSRs. In the context of LIV, distinctive signatures might manifest as temporal disparities between two energy ranges or deviations from the conventional spectra, while also factoring in necessary corrections for photon interactions with the EBL. Studies on the LIV have been conducted by the H.E.S.S. collaboration with observations of PKS 2155-304 (Bolmont et al. 2009) and Mrk 501 (Lorentz and Brun 2017) flares.

3.1.3 Primordial Black Holes

Hypothetical black holes that originated in the early Universe, referred to as **Primordial Black Holes** (PBHs), hold the potential to exhibit distinctive VHE signatures.

Unlike astrophysical black holes that form from the gravitational collapse of massive stars, PBHs were created shortly after the Big Bang within highly dense regions. These enigmatic entities can span a wide range of masses, extending from the Planck mass to thousands of times the mass of our Sun. To investigate the existence of PBHs, one searches for gamma-ray flares characterized by durations ranging from several microseconds to several seconds, which can provide insights into the evaporation process of PBHs with masses approximately around 10^{15} grams. This particular mass reference is chosen because PBHs lighter than this value would have entirely evaporated within the current Universe's timeline (Halzen et al. 1991). Recent work conducted by the H.E.S.S. collaboration has contributed to setting constraints on the rate of PBH evaporation (Aharonian et al. 2023).

3.2 Gamma Rays from Dark Matter Annihilation

3.2.1 Energy-Differential Gamma-Ray Flux

Consider a volume dV with a density ρ_{DM}, at a distance D from the observer, the rate of annihilation per volume and per time writes:

$$\frac{dN}{dVdt} = \frac{n_{DM}^2}{2}\langle\sigma v\rangle = \frac{\rho_{DM}^2}{2\,m_{DM}^2}\langle\sigma v\rangle. \tag{3.1}$$

With dN/dE the spectrum produced per annihilation, the spectrum per annihilation/volume/time is:

$$\frac{dN}{dEdVdt} = \frac{\rho_{DM}^2}{2\,m_{DM}^2}\langle\sigma v\rangle\frac{dN}{dE}. \tag{3.2}$$

The spectrum per unit of time observed in a detector area of dA is therefore:

$$\frac{dN_{obs}}{dEdt} = \frac{\rho_{DM}^2}{2\,m_{DM}^2}\langle\sigma v\rangle\frac{dN}{dE}dV\frac{dA}{4\pi D^2}. \tag{3.3}$$

Integrating over the volume dV in spherical polar coordinates ($dV = R^2dRd\Omega$), the observed spectrum per unit of time observed per detector area dA is:

$$\frac{dN_{obs}}{dEdAdt} = \frac{1}{8\,m_{DM}^2}\langle\sigma v\rangle\frac{dN}{dE}\int\rho_{DM}^2dRd\Omega \tag{3.4}$$

The integral corresponds to the J-factor:

$$J(\Delta\Omega) = \int\rho_{DM}^2dRd\Omega. \tag{3.5}$$

This term encompasses astrophysical information regarding the distribution of DM surrounding the target. More details about the J-factor are explained later in Sect. 3.4. The gamma-rays resulting from DM annihilation within DM-dense environments have the potential to be detected by Imaging Atmospheric Cherenkov Telescopes (IACTs, see Chap. 5). However, the anticipated gamma-ray flux depends upon the assumptions regarding the annihilation processes and the distribution of DM within the target. Therefore, including this information, the gamma-ray flux can be expressed as:

$$\frac{d\Phi_\gamma}{dE}(E, \Delta\Omega) = \underbrace{\frac{1}{4\pi} \frac{\langle\sigma v\rangle}{2\,m_{DM}^2} \sum_i Br_i \frac{dN_i}{dE}(E)}_{\text{particle physics}} \times \underbrace{J(\Delta\Omega)}_{\text{astrophysics}}. \tag{3.6}$$

The equation consists of two primary components. The first term encapsulates the particle physics information concerning the properties of the DM particles. This includes the DM mass m_{DM}, the thermally averaged velocity-weighted annihilation cross-section $\langle\sigma v\rangle$, the annihilation spectrum dN/dE_i specific to a given channel i, and the corresponding branching ratio Br_i.

To estimate the quantity of gamma-rays detected by the telescope, several factors must be taken into account. These factors include the actual flux of gamma-rays $d\Phi_\gamma/dE$, characteristics of the detector, and observation-related information. Therefore, by performing a convolution of the differential flux with the detector's effective area A_{eff}^γ and energy resolution $G(E'-E)$, integrating over the observation time T_{obs} and energy range ΔE, one can calculate the number of gamma-rays counted as:

$$N_{S,\gamma} = T_{obs} \int_{\Delta E} \frac{d\Phi_\gamma}{dE}(E, \Delta\Omega) A_{eff}^\gamma G(E'-E)dE. \tag{3.7}$$

Instrument response functions—like the effective area and the energy resolution—depend on the observation conditions: zenith angle, *offset* of the observations, and energy. More details on the instrument response functions are provided in Chap. 5.

3.2.2 Targets for Very-High-Energy Gamma-Ray Observations

In the DM search quest, it is imperative to direct observations toward targets featuring highly concentrated DM halos or clumps. These densely populated regions serve as potential sites where relic DM particles may undergo processes such as decay or annihilation, producing detectable gamma-ray signals. A more comprehensive exposition of the anticipated gamma-ray flux stemming from DM annihilation will be provided in Sect. 3.2. This section offers a concise introduction to the DM-dense regions that can be observed in the Universe.

The **Galactic Center**, situated at a distance of approximately 8.127 kpc from the solar system (GRAVITY Collaboration et al. 2018), represents the closest and most promising target for detecting DM signals. The GC region is anticipated to harbor a substantial amount of DM. When assuming a standard Navarro Frenk and White (NFW) (Navarro et al. 1997) density profile, the integrated square density of DM along the line of sight within a 1-degree region surrounding the GC is calculated to be $\log_{10}(J/\text{GeV}^2\text{cm}^{-5}) \simeq 21.0$. The parameter J-factor serves as a measure of the DM distribution within an astrophysical system and plays a fundamental role in quantifying the strength of the DM signal emanating from the observed target. The next section will provide further details regarding the J-factor. Consequently, the GC is expected to yield the most substantial gamma-ray signal from DM. However, the presence of several sources within the vicinity of the GC emitting VHE gamma-rays complicates the modeling of expected background signals. The GC region has been subject to extensive observations conducted by the H.E.S.S. and *Fermi*-LAT experiments (the first one will be discussed more in detail in Chap. 5, whereas the second one will be briefly mentioned in Sec. 4.3.1). These observations have yielded the most stringent constraints on DM annihilation at the moment of the writing. More detailed information about the DM search within the inner halo of the Milky Way using the Inner Galaxy Survey dataset can be found in Abdalla et al. (2022) and will be discussed in Chap. 7. An excess of GeV gamma rays from the GC region has been reported by *Fermi*-LAT, with a potential DM annihilation hypothesis (Lacroix et al. 2016). However, more conventional astrophysical explanations are considered more plausible, particularly in the absence of DM detection in cleaner environments like dwarf galaxies (Cembranos et al. 2013). Studies have also been linking the TeV gamma-ray flux observed by H.E.S.S. towards Sgr A* to potential DM signals (Cembranos et al. 2013). An observation by *Fermi*-LAT suggested a hint of a DM signal with a DM mass of 130 GeV near the GC (Su and Finkbeiner 2012), but this claim was subsequently disproved by H.E.S.S. (Abdalla et al. 2016).

Dark matter subhalos are structures of various sizes that are predicted to exist within the main DM halos, as indicated by cosmological simulations. When viewed from Earth, it becomes possible to detect gamma rays emanating from those subhalos situated within the Milky Way. Small subhalos often lack the substantial gravitational potential required for amassing sufficient matter and initiating star formation. Consequently, they might not emit gamma rays except through the signals generated by DM annihilation processes. Some of these subhalos could be close to Earth, boasting a significant DM density that renders them potentially detectable. However, their exact positions remain entirely unknown. Consequently, adopting a strategy based on pointing observations is far from ideal. Extensive wide-field observations conducted with *Fermi*-LAT have uncovered a population of sources that lack counterparts at other wavelengths, referred to as the unidentified Fermi objects (UFOs). These UFOs represent promising candidates for DM subhalos (Kamionkowski et al. 2010), and they have been observed by H.E.S.S. and used to obtain constraints on DM parameters (Abdalla et al. 2021).

Dwarf galaxies within the Local Group—satellites of the Milky Way—represent some of the most DM-dominated entities in the Universe. These galaxies are situated relatively close to Earth, ranging from 25 to 250 kpc away. Unlike other galaxies, these galaxies may initiate star formation and are nearly devoid of gas, rendering them an ideal and pristine environment for gamma-ray observations. The absence of astrophysical complications makes it straightforward to attribute gamma-ray emissions to DM annihilation processes within them. Furthermore, their proximity to Earth facilitates observations and results in an expected substantial signal compared to other potential targets, such as galaxy clusters. The dwarf galaxies J-factors, integrated in a region of 0.5°, are of the order of $\log_{10}(J/\mathrm{GeV}^2\mathrm{cm}^{-5}) \simeq 18.0 - 19.0$. Thus, dSphs are very promising targets for the unambiguous detection of DM signals. They have been targeted by observations with IACTs. Notably, DM searches with H.E.S.S. have been carried out toward a selection of dwarf galaxies, and the results are detailed in Abdallah et al. (2020).

Galaxy clusters are the most massive systems in the Universe, primarily governed by the gravitational influence of DM. These clusters are composed of approximately 80% DM (Jeltema et al. 2009). However, they are located at considerable distances from the Solar system. The J-factor, integrated within a 1-degree region, typically falls within the range of $\log_{10}(J/\mathrm{GeV}^2\mathrm{cm}^{-5}) \simeq 16.0 - 17.0$ for galaxy clusters. Despite their distance, observations of galaxy clusters have been employed to derive constraints that are orders of magnitude fainter than those obtained from GC observations. Galaxy clusters hold great potential for detecting DM decay processes (Acciari et al. 2018). Due to their immense mass, which can be $10^{14} - 10^{15}$ times that of the Sun, they provide substantial volumes for more efficient searches. The decay of DM within these clusters yields electrons and positrons, which undergo inverse Compton scattering and lose energy rapidly. As a result, they emit gamma-rays before escaping, making galaxy clusters an intriguing target.

For a successful detection, a DM signal must be both strong and distinguishable from background emissions. Searching for DM is, therefore, most effective in targets characterized by large DM content, proximity to Earth, and minimal astrophysical background originating from conventional astrophysical sources. The GC, being the nearest to observers, remains the most promising target. However, unambiguous detection is also feasible in other promising targets like subhalos and dwarf galaxy satellites of the Milky Way, as they offer the lowest possible background levels and are relatively close to observers. A sketch of the several aspects that need to be taken into account when choosing the observational targets for the search for DM annihilation signal is shown in Fig. 3.1, superimposed on an all-sky *Fermi*-LAT view at energies larger than 1 GeV. A strong signal from DM annihilation is more likely to be detected from the GC region of the inner GC halo. However, this region also presents uncertainties due to the presence of background. More robust constraints can be derived from observations of dwarf galaxies, at the cost of a weaker expected signal.

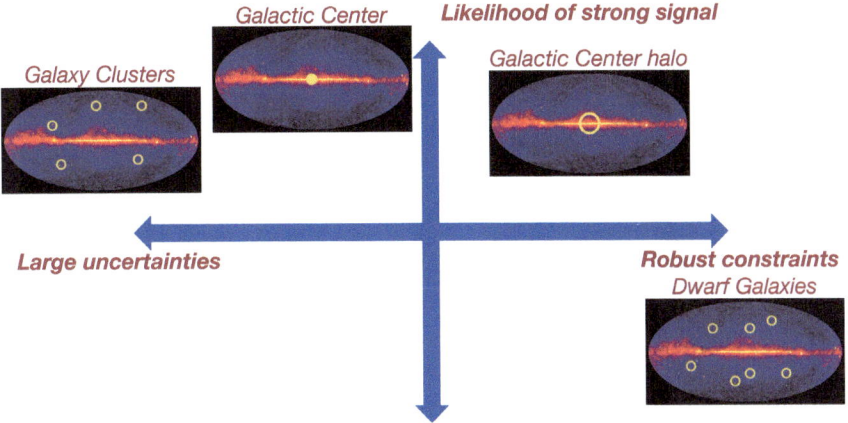

Fig. 3.1 Sketch depicting advantages and disadvantages when observing different targets searching for a DM signal. The indicative position—does not want to be accurate—of the targets is also shown on an all-sky *Fermi*-LAT view, extracted from LAT All-Sky Survey Observations (2024)

3.3 Dark Matter Distribution

3.3.1 Mass Density Profile for the Milky Way

The ability to detect DM signals is directly influenced by the magnitude of the DM density; therefore, making accurate estimations of this density around the target is of utmost importance. When modeling DM halos, various parameterizations are employed, typically falling into two broad categories: cuspy and cored density profiles. In many cases, highly massive galaxies exhibit cuspy profiles, often forming in regions with strong gravitational potentials, such as the central region of the Milky Way, influenced by the presence of the supermassive black hole, Sgr A*.

Two of the most prominent cuspy density profiles in the literature are the Einasto profile (Springel et al. 2008) and the NFW profile (Navarro et al. 1997). The Einasto parameterization is described as follows:

$$\rho_E(r) = \rho_s \exp\left[-\frac{2}{\alpha_s}\left(\left(\frac{r}{r_s}\right)^{\alpha_s} - 1\right)\right] \tag{3.8}$$

and the NFW as:

$$\rho_{NFW}(r) = \rho_s\left(\frac{r}{r_s}\left(1 + \frac{r}{r_s}\right)^2\right)^{-1}. \tag{3.9}$$

The variables in the equations mean: r represents the distance measured from the center of the galaxy, ρ_s stands for the critical density at the location of the Sun, r_s denotes the scale radius where the profile's slope changes, and the steepness of the profile is denoted by α. It's worth noting that the NFW profile goes at infinity at $r = 0$, whereas $\rho_E(0)$ remains finite. Many observations of galaxy rotation curves indicate that the central DM halo has a flat density distribution. In the case of smaller mass galaxies, a central cored profile may be present, and two primary models used for this scenario are the Burkert profile (Burkert 1995) and the isothermal profile (Binney and Tremaine 2008). The Burkert parameterization can be described as follows

$$\rho_B(r) = \rho_s \frac{r_c^3}{(r + r_c)(r^2 + r_c^2)} \tag{3.10}$$

and the isothermal one is written as:

$$\rho_{Iso}(r) = \rho_s \left(1 + \left(\frac{r}{r_c} \right)^2 \right)^{-1}. \tag{3.11}$$

ρ_s is defined as the density inside the core, while r_c represents the radius of the core. From cuspy profiles, cored ones can be obtained by the following modeling:

$$\rho_{E/NFW,core} = \begin{cases} \rho_{E/NFW}(r) & \text{for } r > r_c \\ \rho_{E/NFW}(r_c) & \text{for } r \leq r_c \end{cases} \tag{3.12}$$

These parameterizations have been developed based on N-body simulations and observations of the kinematics of stars and gas. However, it's important to note that these profiles do not incorporate the baryonic component—much more sophisticated simulations are needed in this case (Duffy et al. 2010). Another factor influencing the DM density profile is the interaction with other halos, which can lead to changes through processes like tidal stripping or disruption, potentially tearing apart even the smallest halos (Penarrubia et al. 2008).

In Fig. 3.2, examples of DM density profiles that were cited above are presented. Near the GC, the cuspy profiles are typically three to four orders of magnitude larger than the cored profiles. It's important to recognize that the dynamics of the GC can be effectively reproduced by several different DM profiles, mainly because stars and gas predominantly influence the gravitational potential. The wide-ranging uncertainties associated with the Galactic halo profile significantly impact the search for a DM signal. Consequently, the constraints on the annihilation cross-section obtained can vary by many orders of magnitude depending on the assumed DM density distribution. Table 3.1 reports the parameters adopted for the profiles discussed so far. Einasto, NFW and Einasto 2 parameterizations are used for the results presented in Chap. 7, two of them are also shown in Fig. 3.2. The Burkert and Isothermal parameterizations are only discussed in this section and presented in Fig. 3.2 for comparison. The profiles are normalized such that $\rho(r_\odot) = 0.39$ GeV/cm^3 (Catena and Ullio 2010). As observational estimates of the local density become more precise, any change of $\rho(r_\odot)$ can be propagated to the results by rescaling the DM signal by $(\rho(r_\odot)/(0.39$ GeV/cm^3))^2.

Fig. 3.2 DM distribution in the GC region for examples of cuspy—the Einasto (black line), and NFW (pink line) profiles—and core profiles—the Burkert (blue line) and the Isothermal (dark green) profile. The grey lines indicate the distance at $r_\odot$ from the GC and the value of the local density $\rho(r_\odot) = \rho_\odot$

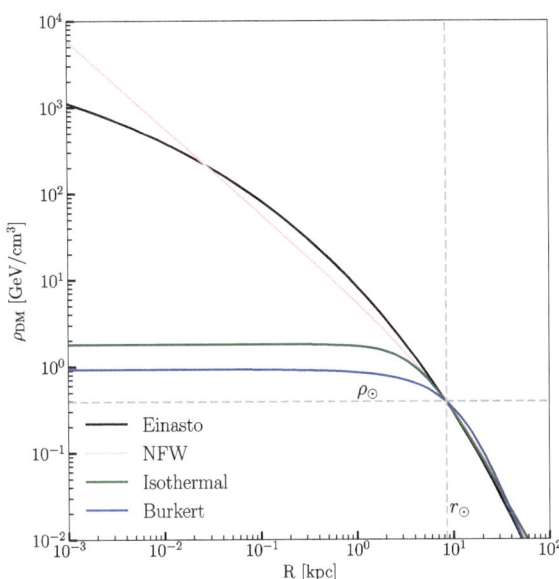

Table 3.1 Parameters of the cuspy and cored profiles used to model the DM distribution. The Einasto and NFW profiles considered here follow Abdallah et al. (2016). An alternative normalization of the Einasto profile (Cirelli et al. 2011) is also used and referred as to "Einasto 2". The cored profile presented in this section follows the Burkert and Isothermal parameterizations

Profiles	Einasto	NFW	Einasto (Cirelli et al. 2011)	Burkert	Isothermal
ρ_s (GeVcm^{-3})	0.079	0.307	0.033	0.712	1.387
r_s/r_c (kpc)	20.0	21.0	28.4	12.67	4.38
α_s	0.17	/	0.17	/	/

Other DM profiles—which will be later used in Chap. 8 for limits derivations—derived with the determination of the Milky Way mass profile exploiting Gaia DR2 measurements of the rotation curve together with in-depth modeling of the baryonic components in the GC (Cautun et al. 2020) are also considered. One of the profiles inferred for the DM distribution used in (Cautun et al. 2020) shows evidence of being contracted by the presence of baryons. In particular, this profile is contracted with respect to the standard NFW profile parameterized in Eq. (3.9). The resulting model provided by the authors is non-parametric, although given its contracted nature it will be referred to as a contracted NFW (cNFW) from now on. Following this approach, it is not possible to measure the distribution within the inner 1 kpc of the Galaxy, and so to be conservative, the density is assumed as cored within this radius. Specifically, $\rho_{cNFW}(r) = \rho_{cNFW}(r_c)$ for $r \leq r_c = 1$ kpc is considered.

The J-factor is typically computed using density profiles resulting from N-body simulations that ignore baryonic effects. However, substantial deviations from the spatial distribution predicted by DM-only simulations have been revealed by realistic hydrodynamic simulations of MW-like galaxies (Abazajian et al. 2014). In particular kiloparsec-sized DM cores—due to the impact of the Galactic bar and stellar feedback—are predicted by some hydrodynamical simulations, including baryonic effects (Chan et al. 2015). The authors of McKeown et al. (2022) computed high-resolution J-factor maps using hydrodynamical FIRE-2 zoom-in simulations. The FIRE-2 simulations include radiative heating and cooling for gas, stellar feedback from OB stars, type Ia and type II supernovae, radiation pressure, and star formation effects.[1] The central DM velocity dispersion was significantly amplified in FIRE-2 (by factors of $\sim$2.5 $-$ 4) compared to DM-only simulations (McKeown et al. 2022). The J-factor maps were computed for s, p, and d-wave DM models (McKeown et al. 2022). For velocity-dependent models, such as the p-wave and d-wave DM scenarios, the J-factor is generalized to encompass the velocity distribution of DM particles. A selection of 12 pairs of simulations (for each FIRE-2 simulation, there is a DM-only one) has been extracted by the authors of McKeown et al. (2022) for comparisons with the Milky Way. These simulations were found to be good candidates for comparison with the Milky Way and used to generate synthetic surveys resembling Gaia DR2 in data structure, magnitude limits, and observational error (Sanderson et al. 2020). For each selected simulation, the stellar mass was found to agree with the Milky Way one of $M_\star = (3 - 11) \times 10^{10}$ M$_\odot$. Also virial masses of the halos in these simulations were found within the expectations from the Milky Way $M_{\mathrm{vir}} = (0.9 - 1.8) \times 10^{12}$ M$_\odot$. The averaged virial mass from the FIRE-2 simulations is in very good agreement with the value inferred by Gaia DR2 of $M_{200}^{\mathrm{total}} = 1.08^{+0.20}_{-0.14} \times 10^{12}$ M$_\odot$ (Cautun et al. 2020). The minimum and maximum J-factors were selected for later representations in this chapter. The FIRE-2 simulations reached a minimum spatial resolution of $\sim$ 400 pc, and their study did not consider AGN-like effects from Sgr A*. For later representations, the J-factor profile has been extrapolated to the central supermassive black hole using a linear approximation. From this, an almost flat density profile in the Galaxy's inner $\sim$ 400 pc is implied. This is considered as a likely conservative approach, as the AGN feedback is expected to cause an even more significant enhancement in DM velocities in the vicinity of Sgr A* (Johnson et al. 2019; McKeown et al. 2022). For distance below 400 pc, Sgr A* feedback may impact DM velocities. However, given Sgr A* low mass compared to the expectation from the black hole mass—velocity dispersion relationship found in elliptical and bulge galaxies (Kormendy and Ho 2013), no significant decrease of the velocity would be expected.

---

[1] The impact of galaxy formation on the spatial distribution of DM halos is less dominant, increasing in some cases the central DM density and decreasing in others.

3.3.2 Dark Matter Density in the Solar Neighborhood

The determination of the local DM density, usually defined as an averaged over about 100 pc around the Sun, is based on either local measurements of the vertical kinematics of stars near the Sun or global ones that extrapolate the DM density from the rotation curve using Galactic halo shape assumptions, see Read (2014) for a review. Its precise value is particularly important for direct and indirect DM detection experiments.

In the following, the dark matter density in the Solar neighborhood is taken to be $\rho(r_\odot) = 0.39$ GeV/cm^3 (Catena and Ullio 2010). The authors studied the problem of constructing mass models for the Milky Way, concentrating on features regarding the DM halo component and including a variegated sample of dynamical observables for the Galaxy. Their approach assumed spherical symmetry and DM distributed according to either an Einasto or an NFW density profile. The chosen value of the local DM density is also compatible within uncertainties with the values obtained from a recent analysis derived from LAMOST DR5 and Gaia DR2 (Guo et al. 2020). Following Read (2014), Zyla et al. (2020), global methods for the determinations of $\rho(r_\odot)$ lie in the range $(0.2 - 0.6)$ GeV/cm^3. Recent studies of the local DM density from Gaia satellite data yield $(0.4 - 1.5)$ GeV/cm^3, depending on the type of stellar tracers used (Buch et al. 2019). When looking for a DM annihilation signal, any change to $\rho(r_\odot)$ can be trivially propagated to the predicted signal by rescaling by a factor of $(\rho(r_\odot)/(0.39$ GeV/cm$^3))^2$.

3.4 The *J*-Factor

3.4.1 Definition

The term encompassing the astrophysical information regarding the distribution of DM surrounding the target was introduced in Eq. (3.6). This crucial quantity is denoted as the *J*-factor and is computed by integrating the square of the DM density over the line of sight (los) s and the solid angle $\Delta\Omega$. The DM density ρ is assumed spherically symmetric and therefore depends only on the radial coordinate r from the center of the DM halo. The mathematical expression for the *J*-factor is as follows:

$$J(\Delta\Omega) = \int_{\Delta\Omega} \int_{los} \rho^2(s(r, \theta)) ds d\Omega. \tag{3.13}$$

If one is searching for DM decay, the *D*-factor $D = \int_{\Delta\Omega} \int_{los} \rho(r(s, \theta)) ds d\Omega$ substitutes the *J*-factor.

3.4.2 Small-Angle Approximation

Let's consider the density to follow $\rho(r) = \rho_0 r^{-\gamma}$. With $r = \sqrt{s^2 + D^2 - 2Ds\cos\theta}$—$D$ and θ are the distance between the observer and the center of the DM halo and the angle between the direction of observation and the center of the DM halo, respectively—and $x = s/D$, and using Eq. (3.13) the J-factor writes:

$$J(\theta) = \frac{\rho_0^2}{D^{2\gamma-1}} \int_0^\infty \frac{dx}{(1 + x^2 - 2x\cos\theta)^\gamma}, \tag{3.14}$$

Substituting $u = x + \cos\theta$, the integral in the right-hand side of Eq. (3.14) can be solved recursively.[2] In the small-angle approximation, $\theta \ll 1$, and $1/2 < \gamma < 3/2$, one gets:

$$J(\Delta\Omega) \simeq 2\pi^2 \frac{\rho_0^2}{D^{2\gamma-1}} \frac{\Gamma(\gamma - 1/2)}{\Gamma(\gamma)\Gamma(1/2)} \frac{\theta^{3-2\gamma}}{3 - 2\gamma} \tag{3.15}$$

3.4.3 J-Factors for Dwarf Galaxies

Consider the simple example of DM particles annihilating in a spherical dwarf galaxy of radius R, uniform density ρ_D, with $D \gg R$ and located at a distance D. One obtains:

$$J(\Delta\Omega) \simeq \frac{4\pi R^3 \rho_D^2}{3D^2}. \tag{3.16}$$

In order to derive an improved estimate, the Jeans modeling which describes dSphs as an uncompressible system at equilibrium will be used. For a comprehensive derivation of the Jeans equation, see Binney and Tremaine (2008). Assuming spherical symmetry, the Jeans equation writes:

$$\frac{1}{\nu(r)} \frac{d}{dr}\left(\nu(r)\sigma_r(r)\right) + 2\frac{\beta(r)\sigma_r(r)}{r} = -\frac{GM(r)}{r^2} \tag{3.17}$$

$\nu(r)$, $\sigma_r(r)$, and $\beta(r) = 1 - \sigma_t(r)/\sigma_r(r)$ describe the 3-dimensional density, radial velocity dispersion, and orbital anisotropy, respectively, of the stellar component. with $\nu(r)$ is the dSph stellar density. Adopting the Plummer profile to describe the dSph stellar densities, $\nu(r) \propto (1 + 2(r/r_h))^{-5/2}$, where r_h is the half-light radius, and isotropic velocity dispersion ($\beta = 0$) and a constant velocity dispersion.[3] In this case, the radial velocity dispersion equals the line-of-sight velocity dispersion σ_{los},

---

[2] The integral can be rewritten as $I_n = \int_{-1}^\infty du/(u^2 + a^2)^n$, with $a = \sin\theta$. One gets $I_n \simeq (1/a^2)I_{n-1}(n - 3/2)/(n - 1)$ that can be recursively solved with $I_1 = \pi/a$ for $\theta \ll 1$.

[3] These assumptions are broadly consistent with available dSphs data. Velocity dispersion profiles are limited to a several values for the faintest dSphs. For the smallest, only the central velocity dispersions are robust.

and Eq. (3.17) gives:

$$M(r) = -\frac{r^2\sigma_r^2}{G\nu}\frac{d\nu}{dr} = 5\left(\frac{2r}{r_h^2}\right)\left(\frac{1}{1+r^2/r_h^2}\right).$$ (3.18)

One then obtains:

$$M(r_h) = \frac{5}{2}\frac{\sigma_{los}^2 r_h}{G}.$$ (3.19)

Let's now assume the DM density $\rho = \rho_0 r^{-\gamma}$, and with $M(<r) = \int_0^r \rho(u)4\pi u^2 du$, one gets: $M(r) = M(r_h)(r/r_h)^{3-\gamma}$. Therefore, the DM density expresses as:

$$\rho(r) = \frac{5}{8\pi}\frac{\sigma_{los}^2(3-\gamma)}{Gr_h^{2-\gamma}}r^{-\gamma}.$$ (3.20)

With $r = \sqrt{s^2 + D^2 - 2D\,s\,\cos\theta}$ and $x = s/D$, this expression can be used in Eq. (3.13) to obtain:

$$J(\theta) = \frac{25}{64\pi^2}\frac{\sigma_{los}^4(3-\gamma)^2}{G^2 r_h^{4-2\gamma}}\frac{1}{D^{2\gamma}}\int\frac{dx}{(1+x^2-2\,x\,\cos\theta)^\gamma}.$$ (3.21)

Using the result derived in the small angle approximation, one gets:

$$J(\Delta\Omega) \simeq \frac{25}{64}\frac{\sigma_{los}^4}{G^2}\frac{D^{1-2\gamma}}{R_h^{4-2\gamma}}\frac{(3-\gamma)^2}{3-2\gamma}\frac{2\Gamma(\gamma-1/2)}{\Gamma(\gamma)\Gamma(1/2)}\theta^{3-2\gamma}$$ (3.22)

$$J(\Delta\Omega) \simeq \frac{25}{64}\frac{\sigma_{los}^4}{G^2}\frac{1}{D^2 R_h}\left(\frac{D\theta}{R_h}\right)^{3-2\gamma}\frac{(3-\gamma)^2}{3-2\gamma}\frac{2\Gamma(\gamma-1/2)}{\Gamma(\gamma)\Gamma(1/2)}.$$ (3.23)

For the Reticulum II ultra-faint dSph with $D = 32$ kpc, $R_h = 15$ pc and an averaged velocity dispersion $\sigma_{los} = 5$ km s^{-1}, an integration angle of 0.5°, and $\gamma = 1$, $\log_{10}(J/\text{GeV}^2\text{cm}^{-5})(< 0.5°) = 18.7$.

Figure 3.3 shows the *J*-factor computed as a function of the angular radius θ for the ultra-faint dwarf spheroidal galaxy Reticulum II and the irregular galaxy Wolf-Lundmark-Melotte (WLM) on the left and right panels, respectively. The WLM galaxy is one of the most promising galaxies for DM searches, since it offers one of the largest *J*-factor among this type of objects (Gammaldi et al. 2018). The authors of Abdallah et al. (2021) considered a cored DM density profile accounting for the galaxy's star-formation history. Compared to dSphs, irregular galaxies have lower *J*-factor, but the knowledge on *J* for dSphs is limited by the number of spectroscopic measurements of individual stars. In comparison, dwarf irregular galaxies benefit from the numerous measurements of the gas tracer, therefore, smaller uncertainties on the *J*-factor are obtained.

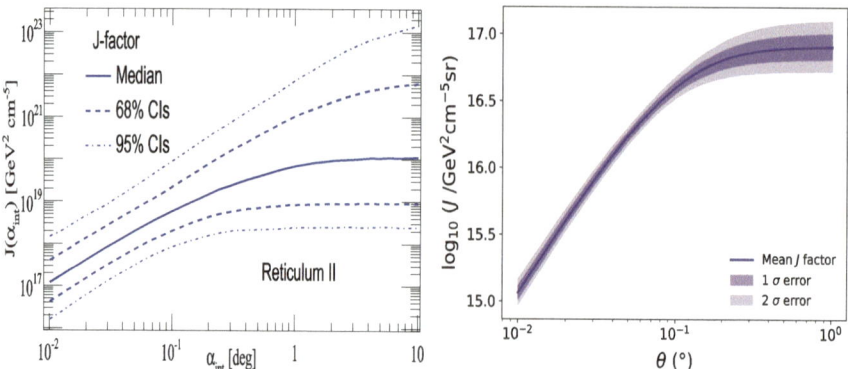

Fig. 3.3 J-factor as function of the angular radius θ computed for the ultra-faint dwarf spheroidal galaxy Reticulum II (left) and irregular galaxy WLM (right). The mean values of the J-factor are shown as the solid line. The 1 and 2σ uncertainty are shown as dashed and dashed-dotted lines on the left panel and bands on the right panel. Figures extracted from Bonnivard et al. (2015) and Abdallah et al. (2021), respectively

3.4.4 J-Factors for the Galactic Center

Figure 3.4 shows the total integrated J-factors $J(<\theta)$ as a function of angular distance θ from the GC. The left panel shows the J-factors computed for the NFW and cNFW profiles as previously described in Sect. 3.3. The profiles were extracted from Cautun et al. (2020), together with the 1σ error bands. The Einasto profile is superimposed on the same panel for the parameterization described in Sect. 3.3. This panel was extracted from Montanari et al. (2023). The right panel shows the J-factors for p-wave annihilation extracted from 12 pairs of cosmological simulations—FIRE-2 including baryon feedback processes and its DM-only companion—as discussed in Sect. 3.3 and extracted from McKeown et al. (2022). For the 12 simulations, the mean, minimum, and maximum J-factors were computed, as displayed in the Figure. This panel was extracted from Montanari et al. (2023).

3.5 Expected Signals from the Inner Milky Way

3.5.1 Annihilation Spectrum

The DM annihilation can lead to various final states of particles. This is contingent upon the DM particle mass being large enough to generate them during the tree-level annihilation process. These final state particles can include leptons, quarks, or bosons. Subsequently, gamma-rays can emerge from the decay or hadronization of

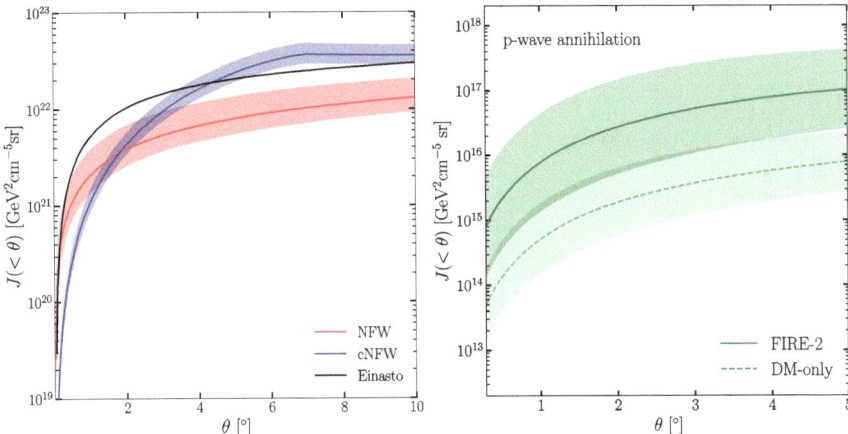

Fig. 3.4 Integrated J-factor (in $GeV^2cm^{-5}sr$) versus angular distance θ (in °) from the Galactic Center for s-wave (left panel) and p-wave (right panel) annihilation, respectively. The 1σ error bands are displayed for the NFW and cNFW profiles. The left panel shows J-factors computed for the Einasto (black), NFW (red), and cNFW (blue) profiles. The right panels shows J-factors for FIRE-2 (dark green) and DM-only (light green) simulations. The shaded areas show the extrema obtained in these sets. The left panel has been extracted from Montanari et al. (2023), whereas the right one from Montanari et al. (2023)

these final state particles. This part of the gamma-ray spectrum generated by such processes is called the "continuum".

Given the assumption that cold DM annihilation occurs at rest, the spectrum exhibits a cutoff at the DM particle mass, denoted as m_{DM}. The spectrum's shape for lower energy ranges varies depending on the specific particles involved in the final state of the annihilation process. Figure 3.5 presents continuum spectra for annihilation in various channels when considering DM particles with a mass of $m_{DM} = 10$ TeV, as computed in Cirelli et al. (2011). Additionally, comparative spectra obtained from a more recent gamma-ray yield study are provided (Bauer et al. 2021). Further details on this comparison will be explored in Chap. 8.

The shape of the spectra varies among different channels. The spectra exhibit a steeper decline near the DM particle mass for leptonic channels, resulting in the spectrum peak being close to this mass. Conversely, the spectra peak typically occurs at approximately $m_{DM}/10$ for bosonic and quark channels. The $\tau^+\tau^-$ channel combines features from both hadronic and leptonic channels, leading to a peak close to $m_{DM}/3$. This particular channel also generates the strongest signal at the peak.

Throughout this book, a 100% branching ratio in each channel will be considered, denoted as the "XX" channel (with X representing potential particles like W, Z, b, t, e, μ, τ, H). However, it's important to note that branching ratios are contingent on factors such as the mass and spin of the particle X and the choice of dark matter particles. As a result, different models can produce varying branching ratios for each candidate and combination. From Fig. 3.5, one can estimate that the number of

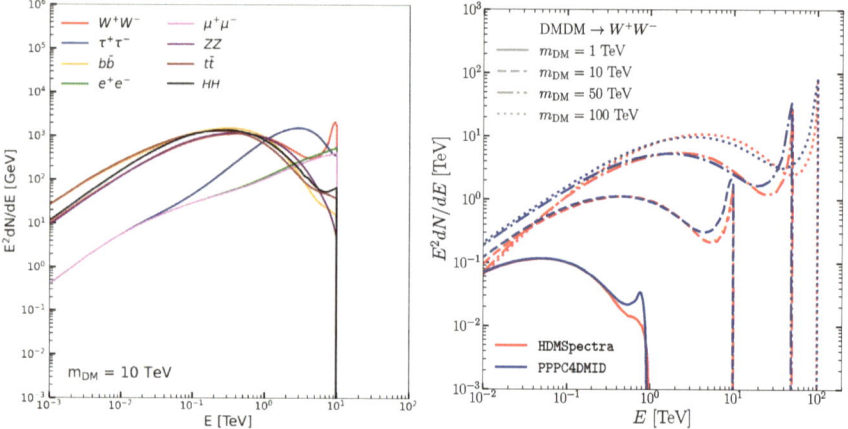

Fig. 3.5 Photons spectra expected from DM particles self-annihilating in the XX channels with X = W, Z, b, t, e, μ, τ, H. *Left panel*: comparison of spectra extracted from PPPC4DMID (Cirelli et al. 2011) for the annihilation channels tested in this work. *Right panel*: comparison of spectra for annihilation of DM in W^+W^- for two gamma-ray yields, PPPC4DMID and HDMSpectra from Bauer et al. (2021). The right panel has been extracted from Montanari et al. (2023)

photons in the continuum channel above the typical energy threshold of 100 GeV is $\int_{0.1\,\text{TeV}} dN/dE\, dE \simeq 0.1$.

The DM annihilation particles can also yield photons, although not at tree-level. Instead, the production of two photons occurs through loops. The cross section of this process is directly related to the square of the electroweak coupling α_{EW}^2 and is consequently suppressed. For WIMPs with masses in the TeV range, the signal strength is considerably smaller, approximately $10^2 - 10^4$ times smaller, compared to the continuum (Profumo 2008).

The photon spectrum from this direct annihilation is often referred to as the "gamma line". This spectrum is characterized by a monoenergetic line shape, resembling a Dirac delta function centered at the mass of the DM particle ($\delta(E - m_{\text{DM}})$). In practical scenarios, instruments with finite resolution may detect a line with some energy spread, which can be modeled using a Gaussian function with a width equivalent to the instrument's energy resolution. The gamma line is the most distinct and sharp signal arising from DM annihilation. However, it also presents the greatest challenge for detection due to its inherently small cross-section and its sensitivity to fluctuations in the observational dataset.

Nevertheless, this channel holds the potential for the most unequivocal DM detection, as no other standard astrophysical process can mimic it.

3.5.2 Flux Estimates for the Continuum and Line Signals

With all the ingredients explained so far, an estimate of what value of the annihilation cross section is needed to reach a $n\sigma$ confidence level for detection can be now made. Considering the DM distribution according to a standard Einasto profile (see Sect. 3.3) with a local DM density of 0.39 GeV/cm^3, Eq. (3.6) yields the following gamma-ray flux in the inner 1 degree around the GC:

$$\Phi_\gamma(> 0.1\,\mathrm{TeV}) \simeq 4 \times 10^{-14}\,\mathrm{cm}^{-2}\mathrm{s}^{-1}\left(\frac{\langle\sigma v\rangle}{2 \times 10^{-26}\,\mathrm{cm}^3\,\mathrm{s}^{-1}}\right)\left(\frac{\int_{0.1\,\mathrm{TeV}}\frac{dN}{dE}dE}{0.1}\right)\left(\frac{m_{\mathrm{DM}}}{1\,\mathrm{TeV}}\right)^{-2}.$$

$$(3.24)$$

For a residual background $d\Phi_{\mathrm{Res.Bkg.}}/dE \simeq 10^{-8}\,\mathrm{TeV}^{-1}\mathrm{cm}^{-2}\mathrm{s}^{-1}(E/1\mathrm{TeV})^{-2.3}$ assumed to be isotropic (see Chap. 7 for more details) and a DM annihilation signal such that $dN/dE = 2\delta(E - m_{\mathrm{DM}})$, one can estimate the value of $\langle\sigma v\rangle_{\gamma\gamma}$ for a detection at $n\sigma$ confidence level. One requests that the number of gamma rays from DM annihilation exceeds the statistical background approximated such that $N_\gamma^{\mathrm{DM}}/\sqrt{N_{\mathrm{Res.Bkg.}}} \geq n$. Assuming an observation time of 100 h, and a constant acceptance of 10^5 m^2 above 100 GeV, one can show that:

$$\langle\sigma v\rangle_{\gamma\gamma} \gtrsim 2n \times 10^{-27}\,\mathrm{cm}^3\,\mathrm{s}^{-1} \times \left(m_{\mathrm{DM}}/1\mathrm{TeV}\right)^2,$$

$$(3.25)$$

for DM masses above 100 GeV at $n\sigma$ confidence level.

Conversely, let us assume DM annihilation for one of the continuum channels. Therefore, let us consider $\int_{0.1\,\mathrm{TeV}} dN/dE\,dE \simeq 0.1$ above an energy threshold of 100 GeV. With the same assumptions made earlier, one can obtain:

$$\langle\sigma v\rangle_{\mathrm{continuum}} \gtrsim 3n \times 10^{-25}\,\mathrm{cm}^3\,\mathrm{s}^{-1} \times \left(m_{\mathrm{DM}}/1\mathrm{TeV}\right)^2,$$

$$(3.26)$$

for DM masses above 100 GeV at $n\sigma$ confidence level.

3.5.3 Astrophysical and Particle Physics Enhancement

Certain particle physics processes can produce additional photons, amplifying the DM signal. Notably, the most significant contributions in these particle physics processes are the Electro-Weak (EW) corrections and the Sommerfeld enhancement. The spectra previously presented already account for the EW corrections. In particular, inverse Compton scattering with ambient radiation in the interstellar medium, such as the CMB, can occur for states containing light leptons as outlined in Cirelli and Panci (2009).

Furthermore, the primary source of astrophysical contribution to the DM signal amplification is attributed to the presence of subhalos.

Electroweak corrections consist in additional radiation due to annihilation into a couple of charged particles of DM particles with a mass larger than the EW scale ($\geq$ 100 GeV) (Ciafaloni et al. 2011). In particle physics, the generation of an additional photon from one of the particles at the interaction vertex is referred to as *final state radiation* (FSR). On the other hand, when virtual exchanged particles produce a photon, this phenomenon is termed *virtual internal bremsstrahlung* (VIB). At larger DM masses, the intensities of these emissions increase. Both FSR and VIB contribute to the appearance of distinct, sharp, line-like features at the end of the spectrum, typically in proximity to the DM particle mass m_{DM}. These notable features are discernible in the spectrum depicted in Fig. 3.5, particularly in the spectra related to the W boson channel.

The Sommerfeld effect is a classical quantum effect happening at low-velocity regimes, when two DM particles in the initial state exchange the interaction mediator many times before the annihilation takes place (Sommerfeld 2006). One can obtain the Sommerfeld enhancement by solving the $l = 0$ Schrödinger equation for the reduced two-body wave function $\Phi(r)$:

$$\left(\frac{1}{m_{DM}} \frac{d^2}{dr^2} - V(r) \right) \Phi(r) = -m_{DM} \beta^2 \Phi(r) . \tag{3.27}$$

The boundary condition $\Phi'(\infty)/\Phi(\infty) = i m_{DM} \beta$ has been considered. The Sommerfeld factor can then be obtained by $S = |\Phi(0)/\Phi(\infty)|^2$. In this scenario, particles interacting through a Yukawa-like potential described as $V(r) = -(\alpha/r) \exp(-m_V r)$ are considered, involving the exchange of a vector boson with mass m_V, where α represents the coupling constant.

This process occurs at non-relativistic velocities and is typically important within DM halos where DM particles move at relative velocities of the order of $\beta = v/c = 10^{-5}$ (with $v = 10$ km s^{-1}). An enhancement factor denoted as $S(\beta, m_{DM}, m_V)$ is applied to the initial value of the thermal relic cross section, denoted as $\langle \sigma v \rangle_0$, resulting in the modified value $\langle \sigma v \rangle = S(\beta, m_{DM}, m_V) \langle \sigma v \rangle_0$. Depending on the relative velocity β, as outlined in Lattanzi and Silk (2009), three distinct regimes can be defined, as illustrated in Fig. 3.6. At large velocities, where $\beta \gg \alpha$, there is no significant enhancement (*i.e.*, $S(\beta, m_{DM}, m_V) = 1$). This regime is considered for scenarios with $\beta = 10^{-1}$. At intermediate velocities, when $\sqrt{\alpha m_V / m_{DM}} \ll \beta \ll \alpha$, the enhancement scales inversely with velocity, such that $S(\beta, m_{DM}, m_V) \simeq \pi \alpha / \beta$, independently of the masses. This regime is illustrated by the green line for $\beta = 10^{-2}$ and does not exhibit resonance behavior. At small velocities, where $\beta \ll \sqrt{\alpha m_V / m_{DM}}$, resonance effects occur due to the presence of bound states. These resonances are exemplified by the yellow, magenta, and purple lines, corresponding to $\beta = 10^{-3} - 10^{-5}$. The degree of enhancement is contingent on the particle masses and scales proportionally to $1/\beta^2$. In DM halos, which typically have very small velocities, the increase in the relic cross section can be substantial, up to a factor of 10^5, due to these resonances. The exact location of the resonances is determined by the masses of the DM and the mediator m_V, signifying the strength

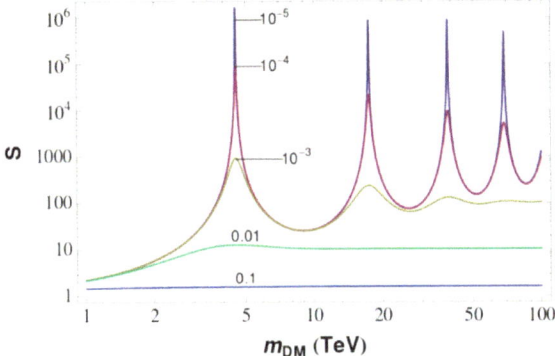

Fig. 3.6 The plot illustrates the Sommerfeld effect's impact on the DM annihilation into the W^+W^- channel, which is mediated by the Z boson. This intensity is depicted as a function of the DM mass. The figure takes into account a range of relative DM velocities spanning from 10^{-1} to 10^{-5} to observe the Sommerfeld effect's behavior across different velocity regimes. Figure extracted from Lattanzi and Silk (2009)

of the coupling between the mediator and the DM particle. Smaller DM masses tend to produce the most significant resonances, while a larger m_V shifts the resonances to higher values of m_{DM}.

Sub-halos are substructures inside the main DM halo predicted by the simulations (see Sect. 3.2.2). The anticipated standard DM signal is calculated based on the even distribution of DM within the host halo. Nevertheless, it has the potential to be enhanced by the presence of subhalos, as discussed in Moliné et al. (2017). Notably, the central density profile or cusp of these smaller halos is steeper than that of larger ones. As a result, the overall J-factor is computed by summing the smooth distribution of the primary halo and the distributions of the subhalos. It's important to note that substructures tend to make a more substantial contribution in the outer regions of the primary halo rather than in the inner regions. Nevertheless, the question of whether the gamma-ray signal is indeed enhanced by the presence of subhalo distributions remains a topic of ongoing debate (Charbonnier et al. 2011).

3.5.3.1 Adiabatic Compression of Dark Matter by Baryon Infall

During infall events such as core collapses (Spitzer 1987), dark matter is compressed toward the center following the adiabatic compression scenario (Zeldovich et al. 1980; Blumenthal et al. 1986). When baryons radiate and contract, DM particles move in a time-dependent potential. Therefore, the DM particle energy is not constant; It can decrease, and the DM particle density can increase. DM can be entrained by the contracting baryons.

Assuming particles moving on circular orbits in a region where $\rho_b(t)$ and $\rho_\chi(t)$ are the baryon and DM densities, respectively, inside the orbit of radius r. The equation of circular motion of χ writes $\ddot{r} + \omega^2(t)r = 0$ with $\omega(t) = \sqrt{4\pi G/3(\rho_b(t) + \rho_\chi(t))}$ given that $v = r\omega$. For a slowly varying ω with time, the adiabatic invariant of the oscillations is $E/\omega \propto A^2\omega$, with A the amplitude of the oscillation, being an adiabatic invariant, i.e., $A^2\omega(t) = A_0^2\omega(0)$. The amplitude of the DM oscillation decreases, and the density increases such that:

$$\frac{\rho_\chi(t)}{\rho_\chi(0)} = \left(\frac{A_0}{A}\right)^3 = \left(\frac{\rho_\chi(t) + \rho_b(t)}{\rho_\chi(0) + \rho_b(0)}\right)^{4/3}. \tag{3.28}$$

Using $x(t) = \rho_b(t)/\rho_\chi(t)$ in Eq. (3.28) gives $x(1 + x)^3 = x(0)(1 + x(0))^3\rho_b(t)/\rho_b(0)$. For $x \ll 1$, one obtains $\rho_\chi(t)/\rho_\chi(0) \simeq 1 + 3(\rho_b(t) - \rho_b(0))/\rho_\chi(0)$. The DM density increases by a factor 3 greater that the baryon density. In case of $x \gg 1$, one can derive $\rho_\chi(t)/\rho_\chi(0) \simeq (\rho_b(t)/(\rho_b(0) + \rho_\chi(0)))^{3/4}$. Such an approach has been applied in the context of DM searches towards Galactic Globular clusters (Wood et al. 2008; Abramowski et al. 2011).

Note, however that this formalism tends to overpredict the compression of the initial dark matter halo (see, for instance, Gnedin et al. (2004)). In particular, this assumes that particles are on circular orbits and does not consider random motions of particles in the halo

3.5.3.2 Adiabatic DM Compression by a Black Hole

Here, an estimate of how the DM distribution reacts to the adiabatic growth of BH in a DM distribution via accretion is shown (Gondolo and Silk 1999). Let's consider an initial distribution of DM particles on circular orbits with $\rho_i(r) \propto r^{-\gamma}$ with a BH growing adiabatically at the center. For a slow process of accretion onto the BH, the angular momentum of each particle is conserved such that $\mathbf{L} = \mathbf{r} \wedge \mathbf{v}$ is constant. Therefore $rv(r)$ is conserved where $v(r) = \sqrt{GM(<r)/r}$ is the circular velocity of the DM particle, $M(<r)$ the total (DM halo and BH) mass enclosed within radius r. From the conservation of the angular momentum for a DM particle with initial and final orbit radii, one gets $r_i M_i(<r_i) = r_f M_f(<r_f)$. The conservation of the DM mass gives $M_i^{DM}(<r_i) = M_f^{DM}(<r_f)$ such that:

$$\int_0^{r_i} \rho_i(r)r^2 dr = \int_0^{r_f} \rho_r(r)r^2 dr. \tag{3.29}$$

With $\rho_f(r) \propto r^{-\gamma_{sp}}$, one gets: $r_i^{3-\gamma} \propto r_f^{3-\gamma_{sp}}$ (i). The total mass in the initial orbit is dominated by the DM halo, i.e., $M_i(<r_i) \simeq M_{DM}(<r_i) \propto r_i^{3-\gamma}$, while in the final state orbit of the DM particle is closer to the BH, i.e., $M_f(<r_f) \simeq M_{BH}$. Therefore, $r_i^{3-\gamma} \propto M_{BH}$. Using the angular momentum conservation, one gets $r_i^{4-\gamma} \propto r_f$ (ii). Putting (i) and (ii) together, one obtains:

$$\gamma_{sp} = \frac{9 - 2\gamma}{4 - \gamma} .$$

(3.30)

Note that this approaches only considers the behaviour of the spatial density. Quinlan et al. (1995) shows however that the slope of the spike depends strongly on the behavior of the initial phase-space distribution.

3.5.3.3 Stellar Heating

If an adiabatic DM spike could have formed, it is inevitably affected by dynamical relaxation from the DM scattering off stars. This process, studied in Gnedin and Primack (2004); Merritt (2004), will smooth the spike and lead to a DM equilibrium profile $\rho(r) \propto r^{-3/2}$. As shown in Vasiliev and Zelnikov (2008), this process depends on the dynamical properties of the stellar core. Following (Binney and Tremaine 2008), the relaxation time expresses as:

$$t_r = \frac{0.34\sigma^3}{G^2 \, m^* \rho^* ln\Lambda} ,$$

(3.31)

with σ the velocity dispersion of the stellar population, m^* and ρ^* the typical stellar mass and density, respectively, and $ln\Lambda \approx 15$ the standard Coulomb logarithm. From Peebles (1980), the radius of the gravitational influence of a BH is given by $r_h = GM_{BH}/\sigma^2$. Following Ferrarese and Ford (2005), it is found that the velocity dispersion is tightly correlated with the central BH mass in a wide range of galaxy masses. From Ferrarese and Ford (2005), one gets: $M_{BH} \simeq 5.7 \times 10^6 M_\odot (\sigma/100 \, kms^{-1})^{4.86}$. One, therefore, obtains that:

$$r_h \simeq 1.4 \, pc \left(\frac{M_{BH}}{4.3 \times 10^6 \, M_\odot} \right)^{0.59} .$$

(3.32)

Inserting the above expression in Eq. (3.31), one obtains:

$$t_r \simeq 1.2 \, Gyr \left(\frac{M_{BH}}{4.3 \times 10^6 \, M_\odot} \right)^{1.4} .$$

(3.33)

For the SMBH of the Milky Way with a mass of $4.3 \times 10^6 M_\odot$, the relaxation time is lower than the age of the universe given by the Hubble time of $1/H_0^{-1} \sim 10$ Gyr. The stellar heating of the DM spike cannot be excluded. However, as discussed in Merritt (2004), the spike may not have had enough time to relax to the equilibrium profile with a slope of $3/2$.

References

Abazajian, K.N. et al.: Astrophysical and Dark Matter Interpretations of Extended Gamma-ray Emission from the Galactic Center, vol. 90.2, 023526, p. 023526, July 2014. https://doi.org/10.1103/PhysRevD.90.023526

Abdalla, H. et al.: Search for Dark Matter Annihilation signals from unidentified Fermi-LAT objects with H.E.S.S. Astrophys. J. **918.1**, 17, 17 (2021). https://doi.org/10.3847/1538-4357/abff59

Abdalla, H. et al.: Search for dark matter annihilation signals in the H.E.S.S. inner galaxy survey. Phys. Rev. Lett. **129.11**, 111101 (2022). https://doi.org/10.1103/PhysRevLett.129.111101

Abdalla, H. et al.: H.E.S.S. Limits on Linelike Dark Matter Signatures in the 100 GeV to 2 TeV Energy Range Close to the Galactic Center. Phys. Rev. Lett. **117.15**, 151302 (2016). https://doi.org/10.1103/PhysRevLett.117.151302

Abdallah, H. et al.: Search for dark matter annihilation in the Wolf-Lundmark-Melotte dwarf irregular galaxy with H.E.S.S. Phys. Rev. D **103**, 102002 (2021). https://doi.org/10.1103/PhysRevD.103.102002. https://link.aps.org/doi/10.1103/PhysRevD.103.102002

Abdallah, H. et al.: Search for dark matter annihilations towards the inner Galactic halo from 10 years of observations with H.E.S.S. Phys. Rev. Lett. **117.11**, 111301 (2016). https://doi.org/10.1103/PhysRevLett.117.111301

Abdallah, H. et al.: Search for dark matter signals towards a selection of recently detected DES dwarf galaxy satellites of the Milky Way with H.E.S.S. Phys. Rev. D **102.6**, 062001 (2020). https://doi.org/10.1103/PhysRevD.102.062001

Abramowski, A. et al.: Acceleration of petaelectronvolt protons in the Galactic Centre. Nature **531**, 476 (2016). https://doi.org/10.1038/nature17147

Abramowski, A. et al.: H.E.S.S. observations of the globular clusters NGC 6388 and M 15 and search for a Dark Matter signal. Astrophys. J. **735**, 12 (2011). https://doi.org/10.1088/0004-637X/735/1/12. arXiv: 1104.2548 [astro-ph.HE]

Acciari, V.A., et al.: Constraining Dark Matter lifetime with a deep gamma-ray survey of the Perseus Galaxy Cluster with MAGIC. Phys. Dark Univ. **22**, 38–47 (2018). https://doi.org/10.1016/j.dark.2018.08.002

Aharonian, F. et al.: Search for the evaporation of primordial black holes with H.E.S.S, vol. 2023.4, 040, p. 040, Apr. 2023. https://doi.org/10.1088/1475-7516/2023/04/040

Bauer, C.W., Rodd, N.L., Webbe, B.R.: Dark matter spectra from the electroweak to the Planck scale. JHEP **06**, 121 (2021). https://doi.org/10.1007/JHEP06(2021)121

Binney, J., Tremaine, S.: Galactic Dynamics, 2nd ed. (2008)

Binney, J., Tremaine, S.: Galactic Dynamics. 2nd ed. (2008)

Blumenthal, G.R. et al.: Contraction of dark matter galactic halos due to Baryonic Infall. Astrophys. J. **301**, 27 (1986). https://doi.org/10.1086/163867

Bolmont, J. et al.: Search for Lorentz Invariance Violation effects with PKS 2155-304 flaring period in 2006 by H.E.S.S. In: 44th Rencontres de Moriond on Very High Energy Phenomena in the Universe, pp. 133–136 (2009). arXiv: 0904.3184 [gr-qc]

Bonnivard, V. et al.: Dark matter annihilation and decay profiles for the Reticulum II dwarf spheroidal galaxy. Astrophys. J. Lett. **808.2**, L36 (2015). https://doi.org/10.1088/2041-8205/808/2/L36. arXiv: 1504.03309 [astro-ph.HE]

Buch, J., Chau, S. (John) Leung, Fan, J.: Using Gaia DR2 to constrain local dark matter density and thin dark disk. JCAP **04**, 026 (2019). https://doi.org/10.1088/1475-7516/2019/04/026. arXiv: 1808.05603 [astro-ph.GA]

Burkert, A.: The structure of dark matter halos in dwarf galaxies. Astrophys. J. Lett. **447**, L25 (1995). https://doi.org/10.1086/309560

Cao, Z. et al.: Ultrahigh-energy photons up to 1.4 petaelectronvolts from 12 γ-ray Galactic sources. Nature **594.7861**, 33–36 (2021). https://doi.org/10.1038/s41586-021-03498-z

Catena, R., Ullio, P.: A novel determination of the local dark matter density. J. Cosmol. Astropart. Phys. **2010.08**, 004 (2010). https://doi.org/10.1088/1475-7516/2010/08/004. https://dx.doi.org/10.1088/1475-7516/2010/08/004

Cautun, M. et al.: The Milky Way total mass profile as inferred from Gaia DR2. Mon. Not. Roy. Astron. Soc. **494.3**, 4291–4313 (2020). https://doi.org/10.1093/mnras/staa1017

Cembranos, J.A.R., Gammaldi, V., Maroto, A.L.: Spectral study of the HESS J1745-290 gamma-ray source as dark matter signal. JCAP **2013.4**, 051, 051 (2013). doi: https://doi.org/10.1088/1475-7516/2013/04/051

Chan, T.K. et al.: The impact of baryonic physics on the structure of dark matter haloes: the view from the FIRE cosmological simulations. Mon. Not. Roy. Astron. Soc. **454.3**, 2981–3001 (2015). https://doi.org/10.1093/mnras/stv2165

Charbonnier, A. et al.: Dark matter profiles and annihilation in dwarf spheroidal galaxies: prospectives for present and future γ-ray observatories—I. The classical dwarf spheroidal galaxies. Mont. Not. Royal Astr. Soc. **418.3**, 1526–1556 (2011). https://doi.org/10.1111/j.1365-2966.2011.19387.x

Ciafaloni, P. et al.: On the importance of electroweak corrections for majorana dark matter indirect detection. JCAP **06**, 018 (2011). https://doi.org/10.1088/1475-7516/2011/06/018

Cirelli, M. et al.: PPPC 4 DM ID: a poor particle physicist cookbook for dark matter indirect detection. JCAP **1103**, 051 (2011). https://doi.org/10.1088/1475-7516/2012/10/E01,10.1088/1475-7516/2011/03/051

Cirelli, M., Panci, P.: Inverse Compton constraints on the Dark Matter e+e- excesses. Nucl. Phys. B **821**, 399–416 (2009). https://doi.org/10.1016/j.nuclphysb.2009.06.034

Duffy, A.R. et al.: Impact of baryon physics on dark matter structures: a detailed simulation study of halo density profiles. Mont. Not. Royal Astr. Soc. **405.4**, 2161–2178 (2010). https://doi.org/10.1111/j.1365-2966.2010.16613.x

Ferrarese, L., Ford, H.: Supermassive Black Holes in Galactic Nuclei: Past, Present and Future Research, vol. 116.3–4, pp. 523–624, Feb. 2005. https://doi.org/10.1007/s11214-005-3947-6. arXiv: astro-ph/0411247 [astro-ph]

Ferrarese, L., Ford, H.: Supermassive black holes in galactic nuclei: Past, present and future research. Space Sci. Rev. **116**, 523–624 (2005). https://doi.org/10.1007/s11214-005-3947-6. arXiv: astro-ph/0411247

Franceschini, A., Rodighiero, G.: The extragalactic background light revisited and the cosmic photon-photon opacity. Astron. Astrophys. **603**, A34 (2017). https://doi.org/10.1051/0004-6361/201629684

Gammaldi, V., Karukes, E., Salucci, P.: Theoretical predictions for dark matter detection in dwarf irregular galaxies with gamma rays. Phys. Rev. D **98**, 083008 (2018). https://doi.org/10.1103/PhysRevD.98.083008. https://link.aps.org/doi/10.1103/PhysRevD.98.083008

Gnedin, O.Y., Primack, J.R.: Dark matter profile in the galactic center. Phys. Rev. Lett. **93**, 061302 (2004). https://doi.org/10.1103/PhysRevLett.93.061302. https://link.aps.org/doi/10.1103/PhysRevLett.93.061302

Gnedin, O.Y. et al.: Response of dark matter halos to condensation of baryons: cosmological simulations and improved adiabatic contraction model. Astrophys. J. **616**, 16–26 (2004). https://doi.org/10.1086/424914. arXiv: astro-ph/0406247

Gondolo, P., Silk, J.: Dark matter annihilation at the galactic center. Phys. Rev. Lett. **83**, 1719–1722 (1999). https://doi.org/10.1103/PhysRevLett.83.1719. https://link.aps.org/doi/10.1103/PhysRevLett.83.1719

GRAVITY Collaboration et al.: Detection of the gravitational redshift in the orbit of the star S2 near the Galactic centre massive black hole. A&A **615**, L15 (2018). https://doi.org/10.1051/0004-6361/201833718.url: https://doi.org/10.1051/0004-6361/201833718

Guo, R. et al.: Measuring the local dark matter density with LAMOST DR5 and Gaia DR2. Month. Not. R. Astronom. Soc. **495.4**, 4828–4844 (2020). ISSN: 0035-8711. https://doi.org/10.1093/mnras/staa1483. eprint: https://academic.oup.com/mnras/article-pdf/495/4/4828/33378321/staa1483.pdf. https://doi.org/10.1093/mnras/staa1483

Halzen, F. et al.: Gamma rays and energetic particles from primordial black holes. Nature **353.6347**, 807–815 (1991). https://doi.org/10.1038/353807a0

Jeltema, T.E., Kehayias, J., Profumo, S.: Gamma rays from clusters and groups of galaxies: cosmic rays versus dark matter. Phys. Rev. D **80**, 023005 (2009). https://doi.org/10.1103/PhysRevD.80.023005

Johnson, C. et al.: Search for gamma-ray emission from p-wave dark matter annihilation in the Galactic Center. Phys. Rev. D **99.10**, 103007 (2019). doi: https://doi.org/10.1103/PhysRevD.99.103007

Kamionkowski, M., Koushiappas, S.M., Kuhlen, M.: Galactic substructure and dark matter Annihilation in the Milky Way Halo. Phys. Rev. D **81**, 043532 (2010). https://doi.org/10.1103/PhysRevD.81.043532. eprint: 1001.3144

Kormendy, J., Ho, L.C.: Coevolution (Or Not) of supermassive black holes and host galaxies. Ann. Rev. Astron. Astrophys. **51**, 511–653 (2013). https://doi.org/10.1146/annurev-astro-082708-101811

T. Lacroix et al. "Connecting the new H.E.S.S. diffuse emission at the Galactic Center with the Fermi GeV excess: a combination of millisecond pulsars and heavy dark matter? Phys. Rev. D **94.12**, 123008 (2016). https://doi.org/10.1103/PhysRevD.94.123008

LAT All-Sky Survey Observations. https://fermi.gsfc.nasa.gov/ssc/observations/types/allsky/. Accessed 02 Feb. 2024

Lattanzi, M., Silk, J.I.: Can the WIMP annihilation boost factor be boosted by the Sommerfeld enhancement? Phys. Rev. D **79**, 083523 (2009). https://doi.org/10.1103/PhysRevD.79.083523

Lorentz, M., Brun, P.: Limits on Lorentz invariance violation at the Planck energy scale from H.E.S.S. spectral analysis of the blazar Mrk 501. In: Morselli, A., Capone, A., Rodriguez Fernandez, G. (eds.), EPJ Web Conferences, vol. 136, p. 03018 (2017). https://doi.org/10.1051/epjconf/201713603018

McKeown, D. et al.: Amplified J-factors in the galactic centre for velocity-dependent dark matter annihilation in FIRE simulations. Mon. Not. Roy. Astron. Soc. **513.1**, 55–70 (2022). https://doi.org/10.1093/mnras/stac966

Merritt, D.: Evolution of the dark matter distribution at the galactic center. Phys. Rev. Lett. **92**, 201304 (2004). https://doi.org/10.1103/PhysRevLett.92.201304. arXiv: astro-ph/0311594

Moliné, Á. et al.: Characterization of subhalo structural properties and implications for dark matter annihilation signals. Mon. Not. Roy. Astron. Soc. **466.4**, 4974–4990 (2017). https://doi.org/10.1093/mnras/stx026

Montanari, A., Macias, O., Moulin, E.: TeV gamma-ray sensitivity to velocity-dependent dark matter models in the Galactic Center. Phys. Rev. D **108**, 083027 (2023). https://doi.org/10.1103/PhysRevD.108.083027. https://link.aps.org/doi/10.1103/PhysRevD.108.083027

Montanari, A., Moulin, E., Rodd, N.L.: Toward the ultimate reach of current imaging atmospheric Cherenkov telescopes and their sensitivity to TeV dark matter. Phys. Rev. D **107**, 043028 (2023). https://doi.org/10.1103/PhysRevD.107.043028. https://link.aps.org/doi/10.1103/PhysRevD.107.043028

Navarro, J.F., Frenk, C.S., White, S.D.M.: A Universal density profile from hierarchical clustering. Astrophys. J. **490**, 493–508 (1997). https://doi.org/10.1086/304888

Peebles, P.J.E.: The Large-Scale Structure of the Universe (1980)

Penarrubia, J., Navarro, J.F., McConnachie, A.W.: The tidal evolution of local group dwarf spheroidals. Astrophys. J. **673**, 226 (2008). https://doi.org/10.1086/523686

Profumo, S.: Hunting the lightest lightest neutralinos. Phys. Rev. D **78**, 023507 (2008). https://doi.org/10.1103/PhysRevD.78.023507

Quinlan, G.D., Hernquist, L., Sigurdsson, S.: Models of galaxies with central black holes: adiabatic growth in spherical galaxies. Astrophys. J. **440**, 554–564 (1995). https://doi.org/10.1086/175295. arXiv: astro-ph/9407005

Read, J.I.: The local dark matter density. J. Phys. G **41**, 063101 (2014). https://doi.org/10.1088/0954-3899/41/6/063101. arXiv: 1404.1938 [astro-ph.GA]

Sanderson, R.E., Wetzel, A., Loebman, S. et al.: Synthetic Gaia surveys from the FIRE cosmological simulations of Milky Way-mass galaxies. Astrophys. J. Suppl. **246.1**, 6, 6 (2020). https://doi.org/10.3847/1538-4365/ab5b9d

Sommerfeld, A.: Über die Beugung und Bremsung der Elektronen. Annalen der Physik **403**, 257–330 (2006). https://doi.org/10.1002/andp.19314030302. Mar

Spitzer, L.: Dynamical Evolution of Globular Clusters (1987)

Springel, V., White, S.D.M., Frenk, C.S., et al.: A blueprint for detecting supersymmetric dark matter in the Galactic halo. Nature **456**, 73–76 (2008)

Su, M., Finkbeiner, D.P.: Strong Evidence for Gamma-ray Line Emission from the Inner Galaxy, June 2012. arXiv: 1206.1616 [astro-ph.HE]

Vasiliev, E., Zelnikov, M.: Dark matter dynamics in the galactic center. Phys. Rev. D **78**, 083506 (2008). https://doi.org/10.1103/PhysRevD.78.083506. https://link.aps.org/doi/10.1103/PhysRevD.78.083506

Wood, M. et al.: A search for dark matter annihilation with the whipple 10 m telescope. Astrophys. J. **678**, 594–605 (2008). https://doi.org/10.1086/529421.arXiv: 0801.1708 [astro-ph]

Zeldovich, Ya.. B., et al.: Astrophysical constraints on the mass of heavy stable neutral leptons. Sov. J. Nucl. Phys. **31**, 664–669 (1980)

Zyla, P.A. et al.: Review of particle physics. In: PTEP 2020.8 (2020) and 2021 update, p. 083C01. https://doi.org/10.1093/ptep/ptaa104

Part II
Astrophysics with Imaging Atmospheric Cherenkov Telescopes

Chapter 4
A Glimpse of the Sky at TeV Energies

Abstract Some fundamental concepts for the TeV energy astrophysics, together with gamma-ray non-thermal production mechanisms are presented in this chapter. After a short description on the processes that are accelerating cosmic rays in the Universe, the production mechanisms for gamma rays are presented. Some of the experiments and observatories dedicated to the detection of very-high-energy gamma rays are outlined with the related detection technique are briefly depicted. The main astrophysical sources emitting at GeV/TeV energies in the Galactic Center region are presented.

Keywords Cosmic rays acceleration · Gamma-ray production · Gamma-ray experiments · Gamma-ray detection technique · Tev emissions in the galactic center

4.1 Cosmic-Ray Acceleration Processes

Cosmic rays refer (CRs) to high-energy protons, electrons/positrons, and atomic nuclei that travel through space at nearly the speed of light. These particles can originate from various sources, including the Sun, regions beyond our Solar System within our galaxy, or even distant galaxies. When CRs interact with Earth's atmosphere, they give rise to a cascade of secondary particles. Victor Hess and Domenico Pancini, in their pioneering work, ruled out the possibility of these radiations being of terrestrial origin. They made this discovery by observing electroscopes discharging spontaneously in the air. Pancini observed that the radiation diminished in deep waters (Pacini 1912). On the other hand, Hess demonstrated that the radiation increased with increasing altitude (Hess 1912). In 1928, Robert Millikan coined the term *cosmic rays* to describe this extraterrestrial radiation (Millikan and Cameron 1928).

CRs have been observed across a broad energy spectrum and are typically classified into different categories based on their energy levels: (i) low-energy (LE) CRs have energies below 50 MeV, (ii) high-energy (HE) CRs fall within the energy range of 50 MeV to 100 GeV, (iii) very-high-energy (VHE) CRs span energies between

---

A non-thermal emission is a continuous emission of particles whose spectrum is not Maxwellian, *i.e.*, cannot be explained either by thermal bremsstrahlung nor black-body emission.

© The Author(s), under exclusive license to Springer Nature Switzerland AG 2024   69
A. Montanari and E. Moulin, *Searching for Dark Matter with Imaging Atmospheric Cherenkov Telescopes*, https://doi.org/10.1007/978-3-031-66470-0_4

Fig. 4.1 The figure shows the CR spectrum spanning energies from 10^8 to 10^{21} eV. The features known as the *knee* and *ankle* are highlighted. The rate of CRs for different energies are provided. A power-law with index ~ 2.7 is shown in green. Figure extracted from Blandford et al. (2014)

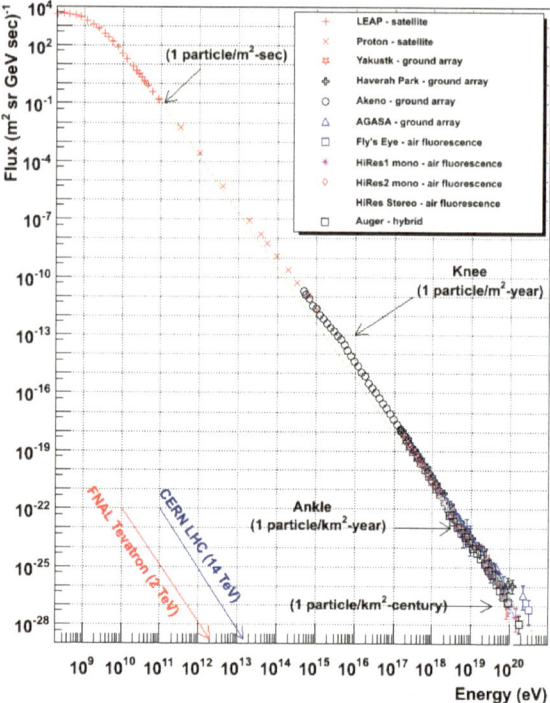

100 GeV and 100 TeV, (iv) ultra-high-energy (UHE) CRs encompass energies above 100 TeV and up to 100 PeV, and (v) extremely-high-energy (EHE) CRs have energies above 100 PeV. Even though this classification is arbitrary, it reflects what is commonly assumed for the categories.

Figure 4.1 displays the measured power-law-like CR spectrum, as discussed in the work by Blandford and Eichler (2014). Several distinct features are observable within this spectrum, including a modulation at low energies due to Solar wind activity, a *knee* at energies around 10^{15-16} eV, marked by a change in spectral index, and an *ankle* at 10^{18-19} eV, characterized by a shift in the spectrum. The rate of CR detection significantly diminishes as energy levels increase. Below the knee, the CR spectrum exhibits a spectral index of approximately 2.7. This index shifts to around 3.3 between the knee and the ankle, and subsequently returns to approximately 2.6 above the ankle. CRs at low and high energies are predominantly believed to originate within the Milky Way (galactic origins). However, CRs with energies exceeding the ankle are likely of extragalactic origin, as they are accelerated to such extreme energy levels by sources beyond our galaxy, such as active galactic nuclei (AGN).

4.1.1 First and Second-Order Fermi Acceleration Processes

The process of accelerating charged particles occurs when they interact with irregularities within a magnetic field. The concept of CR acceleration was introduced by Enrico Fermi in the 1950s, and he developed the second-order Fermi acceleration mechanism, as detailed in his work (Fermi 1949). In this context, it's essential to consider that clouds within the interstellar medium, if perfectly ionized, can be viewed as perfect conductors. These clouds may initially exhibit irregularities in the distribution of magnetic fields if they are magnetized. The second-order Fermi acceleration mechanism involves the following steps: (i) a relativistic particle with an incoming velocity approximately equal to the speed of light (c) enters the cloud, which is itself moving with a velocity (u), (ii) inside the cloud, the particle undergoes random motion and interacts with it, and (iii) due to elastic diffusion on magnetic structures, the particle is reflected with increased energy. In essence, the cloud acts as a magnetic mirror, accelerating particles that approach head-on and decelerating those moving in the opposite direction. This mechanism leads to an average energy gain for the particle, given by $\langle \Delta E/E \rangle = 8/3(u/c)^2 = 8/3 \ \beta^2$ (Fermi 1949). The reason it's termed "second order" is because the energy gain per reflection depends on the square of the particle's velocity β^2. However, it's important to note that the entire observed cosmic ray spectrum cannot be explained solely by this mechanism. It's particularly incapable of accounting for particles accelerated to energies above the GeV range. A linear gain in energy with respect to the cloud's velocity u/c would provide a more efficient explanation of the spectrum, especially considering that $u/c \ll 1$. A visual representation of the second-order Fermi acceleration mechanism is shown in the left panel of Fig. 4.2.

In response to the limitations associated with second-order Fermi acceleration, the scientific community revisited this mechanism in the 1970s and developed the first-order Fermi acceleration mechanism, known as *diffusive shock acceleration*. This mechanism is described in works such as Axford et al. (1977); Bell (1978). Diffusive shock acceleration is fundamentally rooted in the interaction of a relativistic particle with a strong shock wave moving at supersonic velocities. These shock waves propagate through the interstellar medium, and particles are found both in front of and behind the shock. As particles traverse the shock from both directions, they experience isotropic scattering. In the rest frame of the gas, a shock wave approaches

Fig. 4.2 The left and right panels show sketches of the second and first order Fermi acceleration mechanisms, respectively. Figure extracted from Bustamante et al. (2010)

from upstream with a velocity of u_1, while the velocity of the gas beyond the shock is $u = u_1 - u_2$, where u_2 represents the gas's velocity in the shock wave's rest frame. When a relativistic particle crosses the shock upstream at a velocity of v and at an angle θ relative to the shock wave's direction, it gains a small increment of energy, denoted as $\Delta E = E(u/v)\cos\theta$. Subsequently, the particle is scattered to the region behind the shock. At each passage through the shock front, the average energy gain is approximately $\langle \Delta E/E \rangle \simeq u/c$. As the gas downstream is at rest with the gas, it approaches the shock front with a speed of u. Consequently, the same small increase in energy is transmitted to a particle crossing the shock front downstream. This means that a particle crossing the shock multiple times can accumulate numerous energy increases. Diffusive shock acceleration, through repeated crossings of the shock front, allows particles to gain energy efficiently and has become a crucial mechanism in understanding the acceleration of cosmic rays in astrophysical environments. The collisions between particles are always head-on, and there is no energy loss during the crossing. When considering a complete passage from upstream to downstream and then back to upstream, the average energy gain is $\langle \Delta E/E \rangle = 4/3(u/c) = 4/3\,\beta$, meaning it increases linearly with the parameter β. After experiencing n cycles within the acceleration region, the probability that a particle does not escape is given by $P^n = (1 - \langle \Delta E/E \rangle)^n$. The number of particles after n cycles, starting with an initial number of particles N_0, can be calculated as $N_n = N_0 P^n$. Furthermore, the energy of the particles after n cycles is determined as $E_n = E_0(1 + \langle \Delta E/E \rangle)^n = E_0 \epsilon^n$. Consequently, the ratio of the number of particles at this stage to the initial number is described as $N/N_0 = (E/E_0)^{\ln P/\ln \epsilon}$. As a result, the particle spectrum can be approximated as $dN/dE \propto E^{-1+(\ln P/\ln \epsilon)}$. When considering $\ln P/\ln\epsilon \simeq -1$, this approximation leads to a spectral index of approximately 2 at the source. However, as cosmic rays diffuse through the medium, their spectrum is modified, resulting in a softer spectrum with an index ranging from 2.3 to 2.7 when observed far from the accelerators. This diffusion process significantly influences the cosmic ray spectrum. A visual representation of the first-order Fermi acceleration mechanism is given in the right panel of Fig. 4.2.

4.1.2 Astrophysical Accelerators of Cosmic Rays

A **Supernova Remnant** (SNR) is the structure resulting from the explosion of a star in a supernova, which represents the final stage in the life of a highly massive star (Chandrasekhar 1931). After the outer layers are expelled in the supernova explosion, the core undergoes a gravitational collapse, forming a neutron star or a black hole. The specific outcome depends on the star's mass: stars with masses below 10 solar masses become white dwarfs by first shedding the outer layers, those with masses between 10 and $\sim$ 25 solar masses become neutron stars, and those with masses exceeding 30 solar masses usually become black holes.[1] When a star is in the mass

---

[1] The star's metallicity also influences the outcome.

range to become a white dwarf, it gradually accretes mass until it reaches a critical point and undergoes a collapse. The structure of the supernova remnant is composed of the expanding material that was ejected during the explosion, forming a shock front. The shock associated with SNRs can accelerate cosmic rays, which, in turn, can produce gamma-rays (Gabici 2017). It's important to note that the acceleration of cosmic rays within SNRs is powered by approximately 10% of the energy released during the explosion. One of the notable milestones in the field of VHE gamma-ray astronomy was the detection of the first SNR in VHE gamma-rays by H.E.S.S., SNR RXJ1713.7-3946 (Aharonian 2007).

Gamma-Ray Bursts (GRBs) are short and extremely intense bursts of gamma-rays originating from extragalactic processes. These transient phenomena rank among the most luminous events in the Universe. GRBs can arise from the cataclysmic events of either a very massive star's explosion – leading to the formation of a black hole—or the merger of two neutron stars, or even a neutron star and a black hole. The detection of the prompt gamma-rays emitted during GRBs has been achieved by instruments such as Fermi-LAT, capturing gamma rays up to about 50 GeV. The luminosity of these events typically falls within the range of 10^{52} to 10^{54} erg/s. In the keV-MeV energy range, hundreds of GRBs have been identified and analyzed (Narayana Bhat et al. 2016). The MAGIC observatory has managed to detect the prompt emission of GRBs at energies exceeding 300 GeV (MAGIC 2019). Apart from the initial gamma-ray burst, a longer-lived emission known as the *afterglow* is produced. This afterglow results from the interaction between the ejecta from the initial burst and the surrounding interstellar medium. It exhibits a different spectral profile and is detectable across a broader range of wavelengths. H.E.S.S. also detected VHE gamma rays from a GRB afterglow, the brightest one detected so far is GRB221009A (Aharonian et al. 2023). This observation marked a significant advancement in our understanding of these astrophysical events and their high-energy gamma-ray emissions.

Active Galactic Nuclei refer to compact regions situated at the centers of galaxies that exhibit exceptionally high luminosity across various segments of the electromagnetic spectrum. AGNs can be powered by supermassive black holes located at the galaxy's core, with these black holes having masses reaching up to a billion times that of the Sun. In the optical and X-ray ranges, an accretion disk of gas is visible, rotating around these supermassive black holes. Additionally, AGNs emit gamma-rays through highly collimated relativistic jets (Blandford et al. 2019). When these jets are oriented such that they are pointing towards Earth, at an angle smaller than approximately 20°, the object is referred to as a "blazar". A substantial number of blazars have been detected, with some emitting gamma rays in the GeV energy range and a smaller subset producing gamma rays in the TeV range. The primary processes responsible for generating gamma rays in these cases involve inverse Compton scattering (ICS) on synchrotron electrons. The synchrotron self-Compton process is a widely accepted interpretation for the spectra observed in AGNs (Maraschi et al. 1994). The precise role of hadronic processes in the gamma-ray emission from AGNs remains an area of ongoing research, and it is yet to be definitively resolved. The jets are believed to accelerate protons to ultra-high energies, reaching EeV levels.

Among the most prominent examples of blazars with remarkable gamma-ray spectra is the detection of flares from PKS 2155-304 (Aharonian et al. 2007). These observations have significantly advanced our understanding of the high-energy phenomena associated with AGNs and blazars.

Massive stars with core masses falling within the range of 1.5–2.9 solar masses can give rise to the formation of **Pulsars** (PSR) when they undergo supernova explosions (Heger et al. 2003). These pulsars subsequently generate strong stellar winds, and the interaction of these winds with the surrounding medium leads to the creation of **Pulsar Wind Nebulae** (PWN). After the expulsion of the outer layers in a supernova event, the remaining core often becomes a neutron star characterized by high rotational speeds. These neutron stars possess powerful magnetic fields that trap and accelerate charged particles along beams, ultimately ejecting them from the star's poles. Since these ejected beams of particles can be observed periodically from Earth, they are aptly referred to as "pulsars." The typical periodicity of these pulsations is on the order of seconds. Among the gamma-ray-emitting pulsars, the Vela Pulsar is recognized as one of the most energetic (H.E.S.S. Collaboration et al. 2018). The H.E.S.S. observatory has recently detected TeV gamma-ray emissions from the Vela Pulsar (Djannati-Ataı et al. 2017). In the vicinity of the pulsar, charged particles following the magnetic field lines are not ejected as beams but instead corotate with the neutron star. However, at a certain distance from the pulsar, these particles can no longer co-rotate with the star without surpassing the speed of light. Consequently, they escape from the pulsar at this specific distance, termed the *light cylinder* (University of Maryland. 2018). Once these particles are outside the light cylinder, they can be accelerated through shock waves within the surrounding interstellar medium via ICS. This acceleration mechanism effectively generates gamma rays (Bednarek and Bartosik 2003). The Crab Nebula is a prime example of a Pulsar Wind Nebula and stands as one of the most extensively studied PWN in VHE gamma-rays (Aharonian et al. 2006). Moreover, the Crab Nebula has been employed as a standard candle in VHE astrophysical measurements.

4.2   Production Mechanisms and Propagation of Gamma Rays

The production of VHE gamma-rays can occur through two distinct mechanisms: leptonic processes, which involve the acceleration of electrons and positrons, and hadronic processes, which encompass the acceleration of protons and nuclei (Rybicki and Lightman 1986). The relevance of a specific acceleration process varies depending on the energy range considered. The dominant mechanism in the lower energy range, up to tens of keV, is Synchrotron radiation. As the GeV energy range is considered, the primary source of gamma rays is the bremsstrahlung process. In the transition between the GeV and TeV energy ranges, the ICS, along with pion decay, becomes the dominant process. The following sections briefly present each mechanism. For comprehensive reviews, readers are referred to seminal references such as Rybicki and Lightman (1979); Blumenthal and Gould (1970); Longair (1992, 1994).

As it has been explained in Chap. 3, the phenomena involving DM interactions through self-annihilation or decay can produce gamma-rays. The latter can result from primary or secondary processes. The production of gamma-rays through DM annihilation and the expected gamma-ray spectrum have been extensively discussed in Sect. 3.2. This will not be discussed again in what follows.

4.2.1 Leptonic Processes

The interaction of a charged particle with an electromagnetic field gives rise to a phenomenon known as **Synchrotron radiation**. This radiation occurs when a charged particle, typically an electron, experiences acceleration, which results in the particle undergoing a radial motion, effectively spiraling along the lines of the magnetic field. As depicted in the left panel of Fig. 4.3, the process of synchrotron radiation is demonstrated by the emission of electromagnetic radiation when a high-speed electron is bent within a magnetic field. Synchrotron radiation spans a broad spectrum of energies, encompassing wavelengths from radio waves to X-rays, across the electromagnetic spectrum. The photon energy writes as $E_{syn} = 6.7$ eV$(E_e/1$ TeV$)(B/100$ μG$)$ sinα, where α is the angle between the electron trajectory and the magnetic field. Electrons of 10 TeV energy will emit, on average, $\lesssim$ keV photons in a disordered magnetic field of $B_T = 100$ μG.

Bremsstrahlung, often referred to as braking radiation, occurs when a charged particle, typically an electron or a positron, decelerates due to the influence of a Coulomb field associated with an atomic nucleus. During this deceleration process, the incoming particle loses energy, which is then converted into a continuous

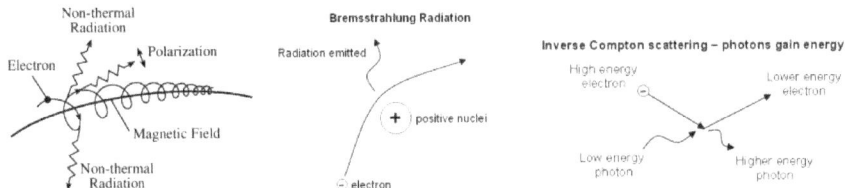

Fig. 4.3 Leptonic mechanisms for producing VHE gamma-rays. *Left panel*: Synchrotron radiation production through the interaction of a charged particle with a magnetic field. Figure extracted from Lang (2010). *Central panel*: Production of gamma rays through the Bremsstrahlung process, when an electron breaks in the electric field of a positively charged nucleus. Figure extracted from Jeff (2005). *Right panel*: VHE gamma-rays production through ICS, when a very energetic electron is scattered against a low energy photon and their energy is exchanged. Figure extracted from Jeff (2005)

spectrum of photons. Notably, Bremsstrahlung becomes the dominant radiative process for electrons/positrons above a few tens of MeV. For muons, the dominance of Bremsstrahlung begins at a higher energy threshold, typically above a few hundreds of GeV. This is because the rate of energy loss through Bremsstrahlung is inversely proportional to the mass of the incoming particle. Consequently, muons lose energy at a slower rate than electrons, leading to the dominance of this process at higher energy levels. When electrons with an energy of E interact with atoms and molecules, this interaction can trigger the production of gamma-rays with frequencies up to $\nu = E/h$, where ν represents the frequency of the emitted gamma-rays, and h is Planck's constant. On average, the energy of the resulting gamma-rays is approximately one-third of the energy of the accelerated particle, denoted as $\langle E \rangle_\gamma = \langle E \rangle_e / 3$. Therefore, electrons in the interstellar medium that have been accelerated to energies in the range of tens of TeV can generate gamma-rays in the TeV energy range. It's important to note that dense environments are more favorable for hosting the Bremsstrahlung process. High-density regions containing many atomic nuclei facilitate the efficient deceleration of the incoming particles, thereby increasing the likelihood of gamma-ray production. Gamma rays produced via Bremsstrahlung will have a spectrum with a spectral index equal to the one of the electron population. A schematic representation of the Bremsstrahlung process is provided in the central panel of Fig. 4.3. More details are discussed at Rybicki and Lightman (1979); Blumenthal and Gould (1970).

The interaction between an accelerated electron and a low-energy photon is referred to as **Inverse Compton Scattering**. In this process, a relativistic electron loses energy while transferring it to photons. The right panel of Fig. 4.3 provides a schematic illustration of this energy exchange. The maximum frequency in the observer's frame during ICS can be approximated as $\nu/\nu_0 \simeq 4\gamma^2$, where γ represents the Lorentz factor of the particle. The resulting photon spectrum is centered around the average frequency, as evidenced by the value of the average frequency $\langle \nu \rangle / \nu_0 \simeq 4/3\gamma^2$. For instance, an electron with an energy of 10 TeV can produce gamma-rays up to about 600 GeV if the target radiation field is the cosmic microwave background. When considering the interaction between a charged particle with energy E_e and mass m and a target photon with energy E in the non-relativistic regime ($E \ll m^2$), the cross section for ICS is approximately equal to the Thompson cross section, denoted as $\sigma_{ICS} = \sigma_T(1 - 2\kappa_0)$, with $\kappa_0 = E/m^2$. The Thompson cross section is roughly $\sigma_T \simeq 6.65 \times 10^{-25}$ cm^2. In this regime, the energy of the scattered photon is approximately $E_\gamma \simeq E^2/m^2$. In the ultra-relativistic Klein-Nishina regime ($E \gg m^2$), the ICS cross section changes to $\sigma_{ICS} = 3/8\sigma_T \kappa_0^{-1} \ln(4\kappa_0)$. Here, the photons produced can have energy levels similar to that of the initial electron. Given a parent particle population of electrons with a spectral distribution described by $dN_e/dE_e \propto E_e^{-\alpha}$, gamma-ray spectra can be characterized as $dN_\gamma/dE_\gamma \propto E_\gamma^{-(1+\alpha)/2}$ in the Thompson regime and $dN_\gamma/dE_\gamma \propto E_\gamma^{-(\alpha+1)} \ln(\kappa_0 + const)$ in the Klein-Nishina regime.

4.2.2 Hadronic Processes

The interaction between accelerated protons and the interstellar gas leads to the creation of neutral pions. These pions subsequently decay into photons (Stecker 1973). In Fig. 4.4, this process is illustrated, depicting how accelerated protons interact with protons in the interstellar medium, generating both charged and neutral pions. Approximately one-third of the produced pions are neutral pions (π^0), while the remaining two-thirds are charged pions (π^+ and π^-) (Kelner et al. 2009). The decay of charged pions gives rise to muons and subsequently neutrinos, whereas neutral pion decay leads to the creation of pairs of gamma-rays through the process $\pi^0 \rightarrow \gamma + \gamma$. This process is characterized by a 98.8% branching ratio and has a relatively short lifetime of $\tau_{\pi^0} = 8.4 \times 10^{-17}$ s. It is worth noting that TeV gamma-rays are most efficiently produced via the decay of neutral pions produced in inelastic pp interactions. The gamma-ray emission traces the gas distribution, which serves as the target for the incident protons. This interaction's energy threshold is approximately $2m_{\pi^0}$, which is approximately 270 MeV. Notably, the gamma-ray spectrum reaches its maximum at around $E_\gamma = m_{\pi^0}/2$, which is approximately 67 MeV, making this a distinct spectral characteristic of this process. The inelastic pp interaction can be approximated as $\sigma^{pp}_{inel} \simeq 34.3 + 1.9 \ln(E_p/1\,\text{TeV})$ mb, noting a logarithmic dependence with the incoming proton energy (Kelner et al. 2009). Starting with an initial spectrum of proton populations characterized by a power-law distribution $dN_p/dE_p \propto E_p^{-\alpha}$, the resulting gamma-ray spectrum follows a power-law with an index of $dN_\gamma/dE_\gamma \propto E_\gamma^{-\alpha+0.1}$ due to a slight energy dependence of the inelastic pp interaction (Kachelriess 2008). Detecting both neutrinos resulting from charged pion decay and gamma-rays from neutral pion decay provides a clear signature of hadronic acceleration by an astrophysical source.

Now that the interesting processes at TeV energies have been introduced, it is interesting to see what is the cooling timescale for a particle population in a turbulent region of the Universe as the inner halo of the Milky Way. Indeed, if one observed cutoff-like features in a measured spectrum of electrons/protons from this region of the sky, this can be due to cooling of the respective particle population.

Fig. 4.4 Interaction of the accelerated protons with the photons of the interstellar medium, and subsequent pion production. Charged pions decay into muons and the corresponding neutrinos, while the neutral ones produce a gamma-ray pair

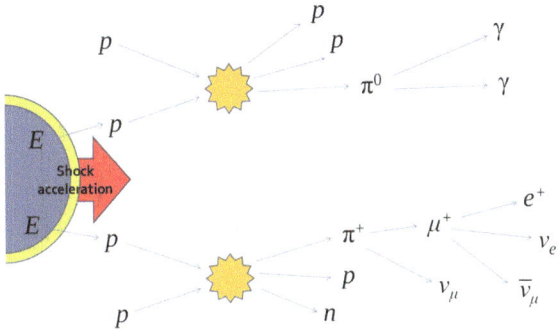

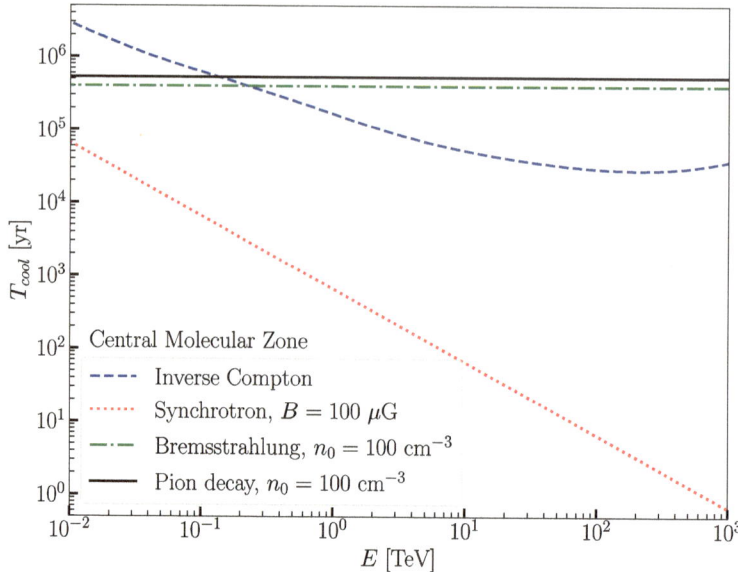

Fig. 4.5 Cooling timescales for electrons and protons as a function of energy. The timescales for electrons include Inverse Compton (dashed blue line), Synchrotron radiation (dotted red line), and Bremsstrahlung emission (dashed-dotted green line). The timescale for protons include Pion decay (solid black line)

As an example, characteristic timescales for the cooling of energetic particles in the dense star formation region at the heart of our Galaxy, the so-called Central Molecular Zone (CMZ), are shown. Therefore, the standard assumptions as if the particles were cooling from the CMZ are made. The characteristic cooling time ($E/\dot{E}$) for electrons include Inverse Compton, Synchrotron radiation, and Bremsstrahlung. The characteristic cooling time for protons is dictated by Pion decay. For Synchrotron cooling, a magnetic field values of ~ 100 μG is assumed, characteristic for the CMZ (Ferrière 2009). Similarly, a gas density of $n_0 \sim 100$ cm^{-3} (Ferrière 2009), determining the cooling timescales for Bremsstrahlung and Pion decay, is adopted (Fig. 4.5).

4.2.3 Propagation of Gamma Rays

VHE gamma rays encounter attenuation when propagating on large distance, the universe becoming opaque at cosmological distances when the photon mean free path is shorter than the distance from the source. The dominant process contributing to the absorption of gamma rays is the creation of electron-positron pairs through the collision of incident VHE photons with extragalactic background photons: $\gamma + \gamma_{background} \rightarrow e^+ + e^-$. The cross-section of this process described by the Bethe-Heitler formula (Heitler 1984). For an isotropic photon background, the cross-section is maximized when the energy of the background photon is $\epsilon \simeq (900 \text{ GeV})/E$ eV (Gould and Schréder 1967).

In general, for VHE photons, this interaction becomes important with optical/infrared photons, making background radiation at such wavelenghts the background component relevant for observations of VHE extragalactic photons. For a 1 TeV energy gamma rays, the cross section is maximized for $\epsilon \simeq 0.5$ eV, *i.e.*, for background photon with near-infrared wavelengths.

This background component is what was introduced earlier as EBL (see Sect. 3.1). As introduced earlier, the degree of attenuation due to EBL absorption is related to the optical depth $\tau(E, z)$. Appreciable modifications of the observed source spectrum—with respect to the spectrum at emission—are due to the energy-dependence of τ. The optical depth increases with energy, therefore, the observed flux gets steeper than the emitted one. One can define the *horizon* (Blanch et al. 2003; Blanch and Martinez 2005) for a photon of energy E as the distance corresponding to the redshift z for which $\tau(E, z) = 1$ from which only 30% of the source flux arrives at the Earth (Fermi LAT studies the EBL 2024).

The direct measurement of the extragalactic infrared radiation field is challenging due to the presence of numerous foreground emissions. The measured attenuation in VHE gamma-rays spectra of blazars place constraints on EBL models, see, for instance, Aharonian et al. (2006).

4.3 Gamma-Ray Experiments and Observatories

Earth's atmosphere is opaque to photons for wavelengths beyond the optical ones. Therefore, high-energy astrophysics requires observations with space-based experiments. However, space-based γ-ray observatories face challenges that do not affect soft X-ray astronomy, for instance.

- High-energy γ-rays cannot be focused. This is not the case for soft X-rays because they can be collected with special mirrors of effective area much greater than that of the detector. For γ-rays, the effective detection area depends on the detector itself, therefore limited to values of the order of $\sim 1 \text{ m}^2$ so that it can fit as a payload for space launchers. Fluxes decrease rapidly for higher γ-ray energies, so space-based observatories are only efficiently sensitive to energies below ~ 100 GeV. For the VHE domain, ground-based detectors are required.
- In the energy range above 100 MeV, γ-rays are primarily identified through their conversion into electron-positron (e^+e^-) pairs in matter, and the incident direction is determined by tracking the paths of the generated electrons. To efficiently detect these γ-rays, it is necessary to utilize converters with a short radiation length. However, when employing such materials, electrons experience significant multiple scattering, which leads to a reduction in angular resolution. This scattering

effect becomes less pronounced as the energy of the particles increases. Nevertheless, even under optimal conditions, the angular resolution typically remains around 0.15°. In comparison, soft X-ray telescopes have the capability to achieve angular resolutions on the order of a few arc seconds.

- Furthermore, both the electron-positron pairs and the gamma rays generated via Bremsstrahlung are captured in a calorimeter, providing the combined energy measurement of the pair. This measurement typically comes with a standard resolution of about 15%.

For the VHE domain above $\sim$ 100 GeV, a different set of techniques is required for observation, and these observations are conducted from the ground. Special telescopes are used to collect the light emitted through the Cherenkov effect resulting from the charged particles in the cascade, initiated by a VHE γ-ray entering the Earth's atmosphere, that move at velocities exceeding the speed of light in the air. These telescopes collect the light even if the electrons and positrons from the cascade do not reach the Earth's surface. The effective detection area is comparable to the area covered by the light pool on the ground, which is around a few 10^4 m^2. This design is well-suited for the very low γ-ray fluxes observed at energies above 100 GeV. An alternative approach is to detect the charged particles within a multi-TeV cascade that reaches the ground in a high-altitude experiment. However, a major challenge is that charged cosmic rays also generate similar cascades in the Earth's atmosphere, creating an extensive background noise compared to genuine γ-ray-induced cascades.

4.3.1 In Space

Experiments positioned on satellites beyond Earth's atmosphere have been designed to detect γ-rays with energies ranging from a few MeV up to approximately 100 GeV. This capability is due to the relatively small size of the gamma-ray detectors on board these satellites, with an effective detection area on the order of $\sim$m^2. These satellite-based detectors offer nearly 100% duty cycles since they continuously observe the cosmos and are not subject to the day-night cycle. They provide modest angular resolution, typically around 0.15 to 0.35°, excellent energy resolution (around 10% or better), and a wide field of view at the order of $\sim$sr.

At the moment of the writing, two recently operational γ-ray telescopes based on satellites are the Fermi Large Area Telescope (*Fermi*-LAT) and AGILE. AGILE, short for Astro-Rivelatore Gamma a Immagini Leggero, was launched in 2007 and de-orbited in February 2024. It features a Gamma-Ray Imaging Detector (GRID) covering the energy range from 30 MeV to 50 GeV, a silicon X-ray detector called SuperAGILE, which operates in the 18 to 60 keV range, and a non-imaging gamma-ray scintillator detector known as Mini-Calorimeter (MCAL), spanning the energy range from 350 keV to 100 MeV. AGILE is also equipped with an anticoincidence

Fig. 4.6 The *Fermi*-LAT space telescope and its instruments: the Large Area Telescope and the Gamma-ray Burst Monitor. Figure extracted from Thompson et al. (2012)

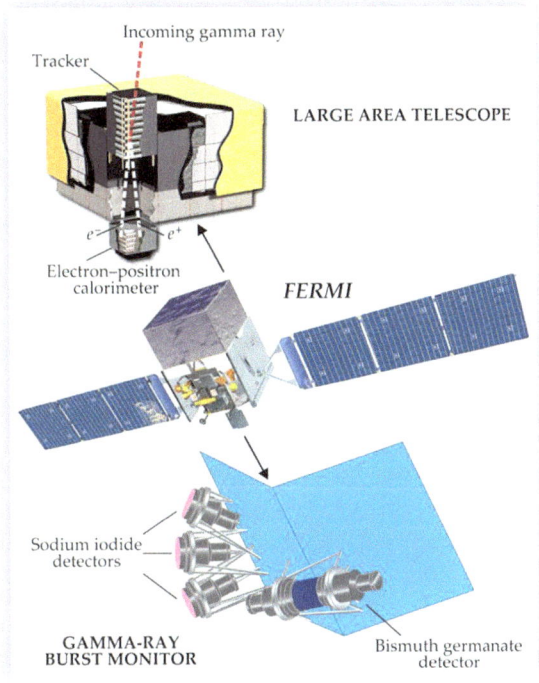

detector serving as a veto (AGILE 2007). *Fermi*-LAT, located on the Fermi Gamma-ray Space Telescope observatory (formerly known as GLAST), offers a wide energy range spanning from 20 MeV to 300 GeV. It provides an energy resolution of less than 10% and maintains a field of view greater than 2 sr. For energies exceeding 10 GeV, Fermi-LAT achieves an angular resolution better than 0.15°. The Large Area Telescope Instrument within *Fermi*-LAT consists of an anticoincidence detector, a tracker, and a calorimeter. The anticoincidence detector helps distinguish γ-rays from background cosmic rays. γ-rays are converted into electron-positron pairs as they interact with tungsten foils in the tracker, and the iodide calorimeter then measures the total energy of the initial γ-ray (Fermi-LAT 2007). The Large Area Telescope Instrument is shown in Fig. 4.6.

4.3.2   On the Ground

4.3.2.1   Water Cherenkov Detectors

Water tank experiments play a crucial role in VHE γ-ray astrophysics. These experiments are designed to detect the secondary charged particles generated by the showers initiated by primary γ-rays as they interact with the Earth's atmosphere. Since these particle showers develop rapidly in the atmosphere and the primary target particles are concentrated in the core of the shower, these experiments are strategically located at high altitudes to ensure they probe this core effectively. As the charged particles within the shower pass through the water tanks, they produce Cherenkov light, which is subsequently detected by photomultipliers (PMTs). This Cherenkov light emission serves as the basis for determining the energy and direction of the incoming primary gamma rays. The spatial distribution of these light signals across the array of water tanks is used to distinguish between γ-rays and CRs. Water tank experiments offer distinct advantages for γ-ray detection at VHE energies, with their highest sensitivity often extending beyond the TeV range. Despite the moderate energy resolution (approximately 50% of the energy), they benefit from good angular resolution ranging from 0.2 to 0.8°, a relatively large field of view of around 2 sr, and an extended operational duty cycle of approximately 90%. At the time of the writing, two prominent observatories employing water tank technology are the High Altitude Water Cherenkov Observatory (HAWC) (HAWC 2022) and the Water Cherenkov Detector Array (WCDA) as part of the Large High Altitude Air Shower Observatory (LHAASO) experiment (Di Sciascio 2016). LHAASO's WCDA facility is particularly notable, featuring a densely packed array of water Cherenkov detectors covering a total area of approximately 78,000 m². This facility is situated at

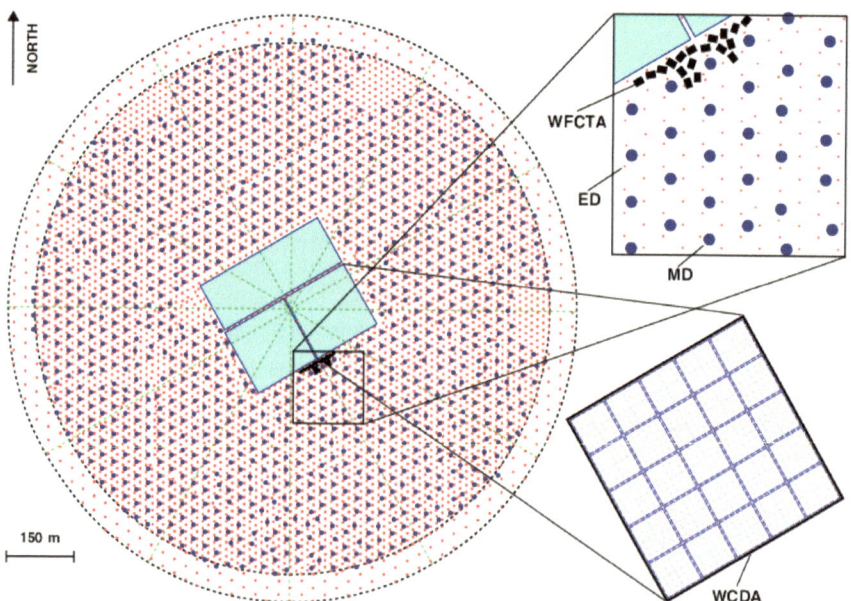

Fig. 4.7 The layout of the LHAASO experiment encompassing all the constituent facilities. In the zoomed portions, the WCDA, some electromagnetic particles (ED) and muon (MD) detectors, and the wide field-of-view air Cherenkov telescopes array (WFCTA) are shown. Figure extracted from Liu (2021)

Table 4.1 Main characteristics of the main WCD arrays: the now discontinued MILAGRO, HAWC, and LHAASO WCDA

Name	Hemisphere	Altitude [m]	Number of water tanks	Number of PMTs	FoV [sr]	E_{thr} [TeV]	Ang. res. at 10 TeV [deg.]	En. res. at 10 TeV as $\Delta E/E$ (%)
MILAGRO	North	2600	1	723	2	0.1	~0.5	~100
HAWC	North	4100	300	1200	2	0.1	~0.2	~50
LHAASO (WCDA)	North	4410	3	6240	2.5	0.05	<0.02	~40

an elevation of 4,410 m in the Sichuan province of China, as depicted in the layout presented in Fig. 4.7. For further details on LHAASO, you can refer to Liu (2021). Some basic information on two currently operating water tank experiments (HAWC and LHAASO WCDA) and the now-ended MILAGRO is reported in Table 4.1.

4.3.2.2 Ground-Based Atmospheric Cherenkov Telescopes

The indirect detection of VHE γ-rays is performed with ground-based imaging atmospheric Cherenkov telescopes (IACTs). These telescopes operate by capturing the Cherenkov light generated when γ-rays interact with atmospheric molecules and induce a cascade of charged particles. IACTs exhibit a sensitivity range covering γ-rays with energies ranging from tens of GeV to approximately 100 TeV. They offer exceptional energy resolution, typically around 10% of the energy, and boast impressive angular resolution, at the order of 0.1°. However, the drawback of IACTs is their relatively short duty cycle, typically hovering around 10–15%, due to daytime inactivity. Moreover, their FoV is somewhat limited, approximately 5° or roughly 10^{-1} sr, necessitating pointed observations. More details about the detection technique with IACTs are provided in Chap. 5.

At the time of the writing, three of the main operational IACTs are the following observatories: H.E.S.S. (The High Energy Stereoscopic System) (H.E.S.S. Collaboration 2002), MAGIC (Major Atmospheric Gamma-ray Imaging Cherenkov Telescope) (MAGIC 2003), and VERITAS (Very Energetic Radiation Imaging Telescope Array System) (VERITAS 2003). The main characteristics of these arrays of telescopes are summarized in Table 4.2. These telescopes present unique characteristics and capabilities that contribute to their effectiveness in VHE γ-ray observations. Notably, the First G-APD Cherenkov Telescope (FACT), situated at the MAGIC site in La Palma, was introduced in 2011 to test innovative technology intended for use in the future Cherenkov Telescope Array (CTA) (FACT 2003). The FACT camera incorporates Geiger-mode avalanche photodiodes (G-APDs) instead of traditional photomultiplier tubes, offering enhanced robustness, lower operating voltage requirements, and improved photon detection efficiency. These G-APDs have also undergone trials

Table 4.2 Main characteristics of three currently operating arrays of IACTs: H.E.S.S., MAGIC and VERITAS

Name	Hemisphere	Altitude [m]	N° of telescopes	Mirror area [m²]	N° of pixels	FoV [deg.]	E_{thr}	Ang. res. at 1 TeV	En. res. at 1 TeV as $\Delta E/E$ (%)
H.E.S.S.	South	1800	4 + 1	108/600	960/2048	5/3.2	0.1/0.03	~0.08	~10
MAGIC	North	2225	2	234	574	3.5	0.06	~0.1	~15
VERITAS	North	1275	4	106	299	3.5	0.1	~0.1	~15

Fig. 4.8 Energy-differential flux sensitivity of current and future VHE gamma-ray observatories. Figure extracted from Hinton and Ruiz-Velasco (2020)

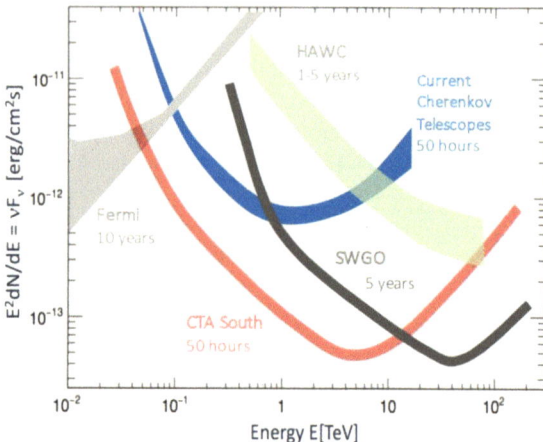

under intense moonlight, aiming to extend the duty cycle for enhanced capabilities in detecting transient emissions, a critical factor in astrophysical observations.

Figure 4.8 shows the energy differential sensitivity of current and forthcoming IACTs as well WCD arrays.

4.3.3   Detection Steps of a High-Energy Gamma Ray

A gamma-ray in the GeV-TeV energy range can first be detected by producing electron-positron pairs as it interacts with the tracker in the *Fermi*-LAT space-based instrument.

Otherwise, the gamma-ray can reach the Earth's atmosphere, where it interacts with an atmospheric nucleus. The gamma-ray undergoes pair production in the vicinity of the nucleus of an atmospheric molecule.

This interaction starts the shower generation, which produces secondary charged particles. These charged particles are detected by water tank experiments—like

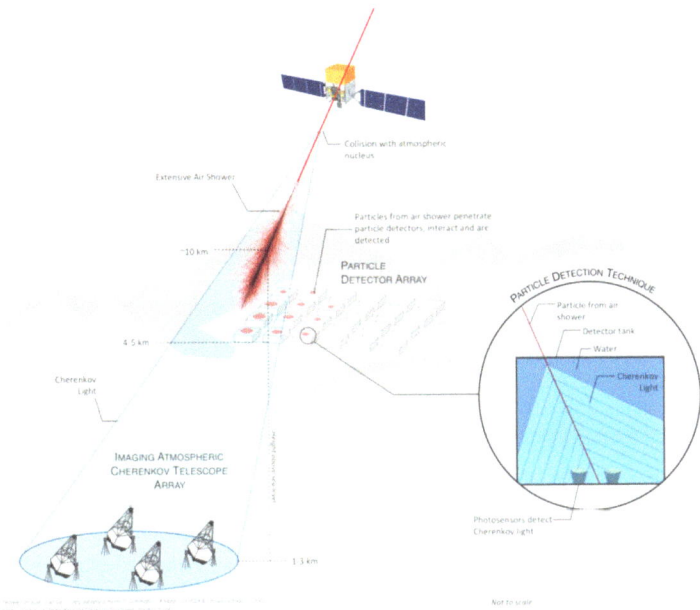

Fig. 4.9 Sketch of the detection of a high energy gamma ray. The first detection is with space-based telescopes like *Fermi*-LAT. Then, the charged particles generated in the shower are detected by water tank experiments like HAWC. Finally the images of the Cherenkov radiation produced by the shower are detected by IACTs like H.E.S.S. Figure adapted from Shower Images (2024)

HAWC—at high altitudes. As the charged particles traverse the water tanks, they produce Cherenkov light, which is subsequently detected by PMTs.

Finally, imaging atmospheric Cherenkov telescopes (IACTs) like H.E.S.S., placed at lower altitudes than water tank experiments, capture images of the very short flash of Cherenkov radiation generated on the ground by the cascade of the relativistic charged particles.

A sketch of this process is shown in Fig. 4.9, considering an hypothetical gamma-ray emitted by the Galactic Center.

4.4 GeV-TeV Gamma-Ray Emissions in the Galactic Center

As the analyses shown later in this book focus on searching for a DM signal in the GC region, this section shows some insights into how the region looks like when observed at high- and very-high energies. A sky map of the region observed by H.E.S.S. is shown in Fig. 4.10. More information about the objects emitting in TeV energies is provided in the next sections. When a DM signal is searched for, these objects must be considered background emissions. More information on how this is

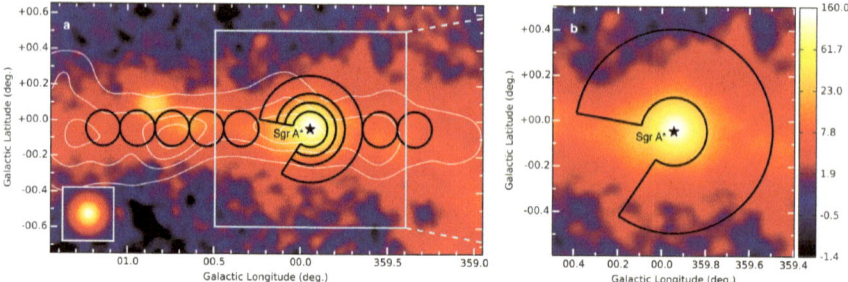

Fig. 4.10 The gamma-ray observation of the GC region conducted by H.E.S.S. The black lines utilized for analyzing the CR energy density in the central zone are juxtaposed with the white lines representing Carbon Monoxide (CO) to Sulfur (CS) line emission. For a more detailed perspective of the inner approximately 70 parsecs, a zoomed-in view is presented on the right side of the figure. Figure extracted from Abramowski et al. (2016)

dealt with is provided in Chap. 7. Sources that have been detected as point-like by H.E.S.S. are easy to exclude from the analysis. However, extended emissions like the Fermi Bubbles cover important parts of the region of the sky observed for the specific purpose of the analyses shown in Chaps. 7 and 8. In this case, one cannot simply exclude these parts of the sky from the analysis because it would reduce the total statistics by half. However, this unaccounted background would still be measured in the region used to search for the signal, contributing to an extended faint excess incompatible with the adopted DM models.

4.4.1   The Central Source HESS J1745-290

The intense TeV emission originating from HESS J1745-290 has been observed by H.E.S.S. in close proximity to the supermassive black hole Sagittarius A* (Baganoff et al. 2003), situated at the gravitational center of the Galaxy. The barycenter of this VHE emission is at $l = 359.94°$, $b = -0.04°$. Sagittarius A*, harboring a mass of 4.31×10^6 $M_\odot$, has been extensively studied across various wavelengths, revealing variability in X-rays and IR (Boyce et al. 2019). However, gamma-ray observations have yet to detect any signs of variability. The composite spectrum, depicted in Fig. 4.11, showcases the VHE emission measured by the H.E.S.S. array, fitting well with an exponential cut-off power law. The cut-off has been computed as $E_{cut} = (10.7 \pm 2.0_{stat} \pm 2.1_{syst})$ TeV, for a power law spectral index of 2.1 and a normalization of 2.55×10^{-12}TeV^{-1}cm^{-2}s^{-1}. Alternatively, a smoothed broken power-law model also yields an excellent fit. The best fit indexes are 2.02 and 2.63, with a break energy computed at 2.57 TeV and a normalization of 2.57×10^{-12}TeV^{-1}cm^{-2}s^{-1}. The submillimeter emission of Sgr A* can be explained by the stochastic acceleration of electrons in the turbulent magnetic field of

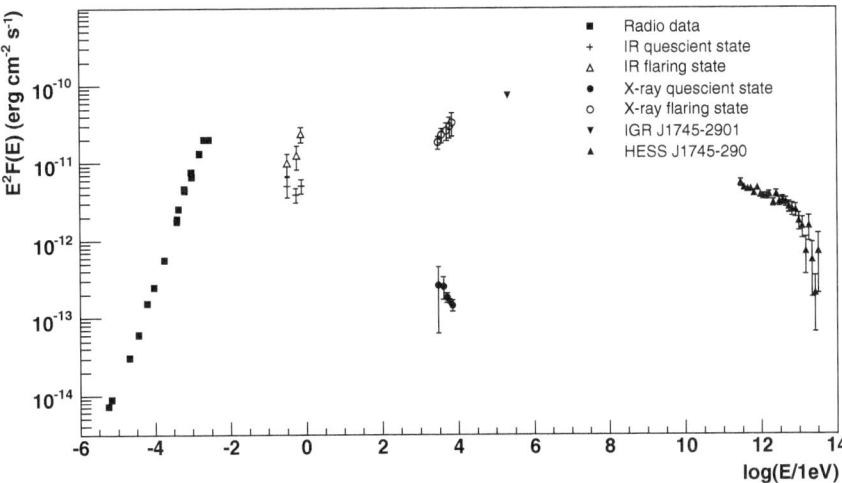

Fig. 4.11 Composite spectrum of Sgr A*. The wide spectral energy distribution is obtained from spectra in radio, IR, X-rays and gamma-rays. Variable flaring states are observed in IR and X-rays. Figure extracted from Aharonian et al. (2009)

the region. This argument also accounts for the IR and X-ray flaring states. Additionally, as charged particles are accreted onto the black hole, escaping protons could undergo acceleration and interact with the interstellar medium in the central star cluster, producing gamma rays (Liu et al. 2006). The cutoff on the proton spectrum can be estimated as $E_{p,cut} \simeq E_{cut}/20$, allowing for the acceleration of protons up to a few hundred TeV. The broken power-law spectrum aligns with models predicting a competition between proton injection and escape. Moreover, IC emissions from electrons accelerated up to approximately 100 TeV in the nearby PWN G359.95-0.04 could also contribute to the TeV emission.

The source HESS J1745-290 could be explained as gamma rays emitted by 10 TeV-ish DM particles annihilating into a combination of $b\bar{b}$ and $\tau^+\tau^-$ channels (see Belikov et al. (2012); Cembranos et al. (2012)). However, this would require a significant overall boost of the DM signals by the contraction of the DM density around Sgr A*. This would result in a so-called dark matter spike. This can arise from the adiabatic growth of the black hole due to the scattering of DM particles with the dense stellar environment of the black hole, or from baryonic infall, as shown in Sect. 3.5.3. From these mechanisms, a factor 100 to 1000 enhancement on the DM annihilation signal can be obtained (Bertone and Merritt 2005). For a DM annihilation signal, distinctive features close to the DM mass like the cutoff (Bringmann et al. 2011) or box-shaped (Ibarra et al. 2012, 2016) spectral features are probably more realistic than the smoking-gun spectral signature in the form of lines at the DM mass. It has also been demonstrated that gamma-ray spectra from DM annihilations into hard channels can prefer the super-exponential power law parameterization rather than the simple exponential cutoff power law (Belikov and Silk 2013). This provides further discrimination against standard astrophysical emissions. Nevertheless, the accuracy of the H.E.S.S. measurements at the highest energy end of the HESS J1745-290 spectrum is insufficient to significantly distinguish between an exponential or a super-exponential cutoff (Lefranc 2016).

4.4.2 The Central Molecular Zone

The dense star formation region situated at the heart of the Galaxy is referred to as the Central Molecular Zone. This area consists of hot gas (Morris and Serabyn 1996; Armillotta et al. 2019) and spans approximately 300 parsecs along the Galactic plane. The CMZ is characterized by intense CS line emissions in the radio spectrum at a wavelength of 1.1 mm (Bally et al. 2010), unveiling the presence of clouds with a cumulative mass of about 60 million $M_\odot$. The average density within the CMZ is notably higher, reaching approximately 100 times the density outside this region, equating to hundreds of atoms per cubic centimeter (Hatchfield et al. 2020).

Key structures within the CMZ include the Sgr A* radio arc complex, Sgr B, Sgr C, and Sgr D, as illustrated in Heywood et al. (2022). Observations across various wavelengths within the inner few degrees of the Galactic Center have unveiled expanding molecular rings, arc structures, and the Galactic Center lobes. These structures play a pivotal role in understanding the dynamic processes occurring at the Galactic Center lobes. While explosive events are hypothesized to be responsible for generating these structures, the specific mechanisms of their formation remain to be elucidated. A more thorough understanding of the morphology, density, and velocity distribution of the underlying gas in the CMZ is crucial for advancing investigations into these phenomena.

4.4.3 The Base of the Fermi Bubbles

Fermi Bubbles (FBs) are extensive bipolar structures with a width of 40° in Galactic longitudes, stretching up to 55° above and below the Galactic plane in latitude. *Fermi*-LAT observations provided insights into their spectrum, revealing a slope of 1.9 and a cutoff at 110 GeV at high latitudes (Ackermann et al. 2017). Recent analyses of VHE FBs emission indicated two components (Herold and Malyshev 2019). The high-latitude FBs spectrum is soft, while the low-latitude component ($|b| < 10°$) appears harder and brighter.

The FBs emission exhibits a photon index of 1.9, differing from the Galactic Diffuse Emission with a photon index of 2.4, making their association less likely. Although the high-latitude FBs spectrum softens considerably above 100 GeV, no similar behavior is observed at low latitudes, and there is no significant hint of a cutoff in the Fermi spectrum. This leaves the possibility open for the observation of a low-latitude FBs component in TeV gamma-rays with H.E.S.S.. The Fermi Bubbles template derived from the hard component of the total Fermi analysis is

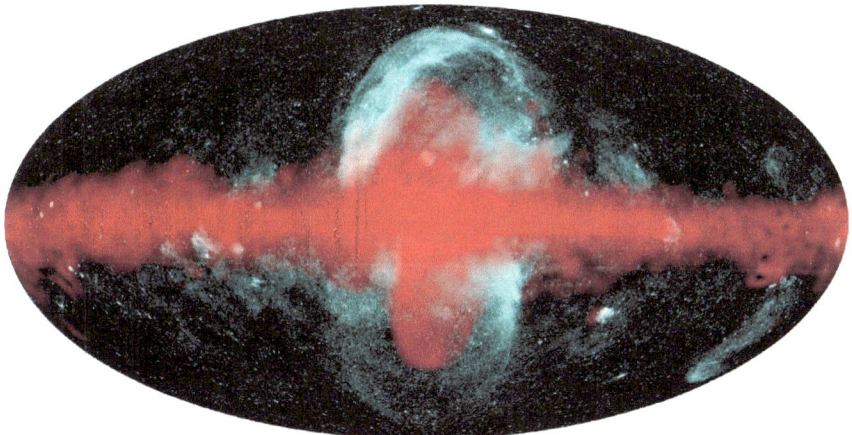

Fig. 4.12 A composite view of Fermi-eROSITA Bubbles. The X-ray observations are shown as the cyan region. The gamma-ray emission, as observed by Fermi, is shown in red. Figure extracted from Predehl et al. (2020)

Fig. 4.13 Derivation of the Fermi Bubbles template from the *Fermi*-LAT analysis. The hard component of the total excess measured in the region is extracted from the analysis to derive the template. This figure was sourced from Ackermann et al. (2017)

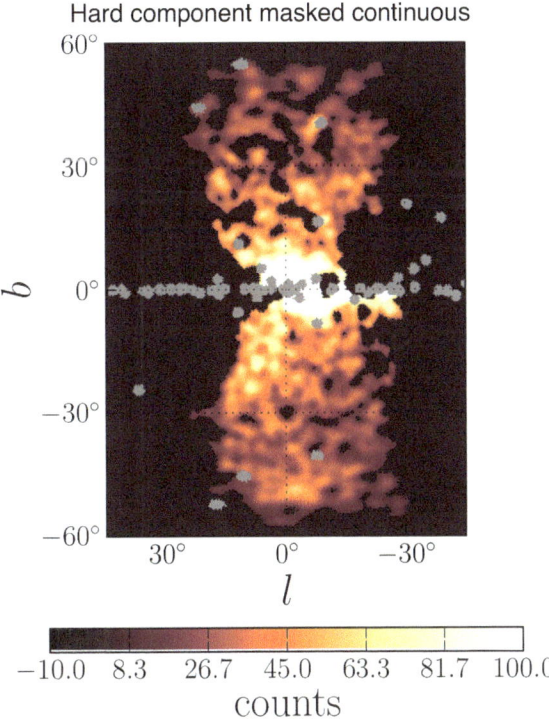

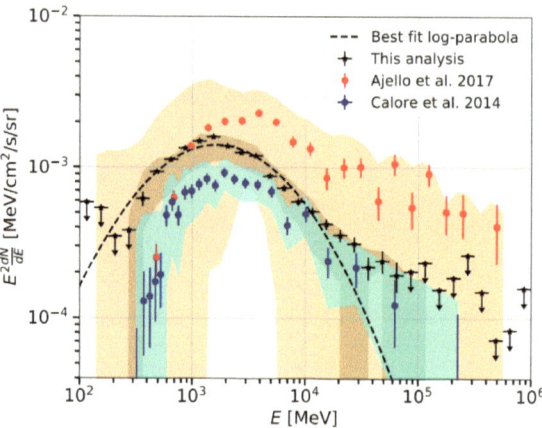

Fig. 4.14 The spectrum of the GCE was reevaluated in Di Mauro (2021) and compared with previous analyses. Various analysis techniques were employed to compute the IEM, contributing to the observed variation represented by the bands in the figure. The best-fit to the GCE spectral energy distribution, achieved with the baseline IEM using a log-parabola function, is prominently displayed. This figure was extracted from Di Mauro (2021)

shown in Fig. 4.13. Spectra extracted from the Fermi article (Ackermann et al. 2017) are depicted in Fig. 4.15, showing the Fermi Bubbles in $|b| < 10°$ and $|b| > 10°$ as teal stars and indigo triangles, respectively.

A structure resembling the Bubbles has been observed by eROSITA (Predehl et al. 2020), showcasing soft-X-ray emitting bubbles extending approximately 14 kpc above and below the plane (Fig. 4.12). These structures enclose the gamma-ray emission detected by the Fermi telescope and appear correlated with the Fermi Bubbles. The production mechanism for these bubbles remains unclear. Their detection, along with the presence of a synchrotron haze, suggests the likelihood of a radio counterpart in scenarios involving leptonic processes. The emission can be reproduced by both leptonic and hadronic processes of gamma-ray production (Ackermann et al. 2017). A recent analysis of the GC region with MeerKAT revealed 430 pc bipolar radio bubbles, likely related to the FBs (Heywood et al. 2022).

4.4.4 The Galactic Center Excess

The central 1-degree region of the GC exhibited a gamma-ray excess (GCE) in GeV *Fermi*-LAT observations, deviating from predictions based on the interstellar emission model (IEM). One of the initial interpretations of this excess was attributed to the annihilation of DM particles with a mass range of 30–50 GeV, assuming a NFW

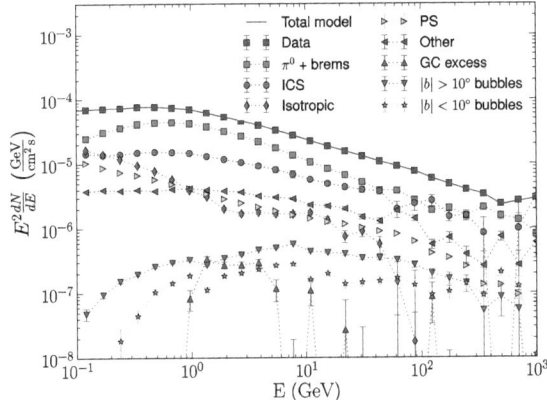

Fig. 4.15 The spectra of the GDE as observed by *Fermi*-LAT and its various components. The data points, represented by blue squares, are accompanied by contributions from different sources: gas-correlated GDE emission (depicted as green squares), ICS radiation (illustrated by orange dots), isotropic background (magenta diamonds), point-like sources (yellow triangles), and the GC GeV excess (shown as green triangles). Additionally, emissions from the Fermi Bubbles are presented for both high latitude ($|b| > 10°$) and low latitude ($|b| < 10°$), displayed as indigo triangles and teal stars, respectively. This figure was sourced from Ackermann et al. (2017)

profile (Navarro et al. 1997), and a relic cross-section on the order of 10^{-26} cm^{3}s^{-1}, as anticipated for thermal production. However, constraints on the relic cross-section for masses up to several hundred GeV had been established from previous *Fermi*-LAT measurements in dwarf galaxies, creating tension with the proposed DM explanation of the GCE. The spectra associated with this hypothesis were found to be strongly dependent on the chosen IEM in a subsequent analysis with updated 6.5 years of Fermi-LAT observations (Ackermann et al. 2017). This analysis also incorporated an additional population of electrons to model the CMZ and considered three distinct point source catalogs.

A more recent investigation reexamined the GCE spectrum using 11 years of Fermi-LAT data (Di Mauro 2021), depicted in Fig. 4.14 from Di Mauro (2021). The spectrum of the GCE from the Fermi analysis is shown in Ackermann et al. (2017), taking into account the interplay with low-latitude emission from the Fermi Bubbles, and is presented as black circles in Fig. 4.15. Between the debates about the nature of the GDE, an alternative hypothesis postulated a population of millisecond pulsars in the Galactic bulge as a plausible explanation. This pulsar population could be shaped by non-spherically symmetric stellar density distributions within the Galactic bulge (Macias et al. 2021).

References

Abramowski, A., et al.: Acceleration of petaelectronvolt protons in the galactic centre. Nature **531**, 476 (2016). https://doi.org/10.1038/nature17147

Ackermann, M., et al.: The fermi galactic Center GeV excess and implications for dark matter. Astrophys. J. **840**(1), 43 (2017). https://doi.org/10.3847/1538-4357/aa6cab

AGILE.: AGILE - Astro-rivelatore gamma a immagini leggero (2007). http://agile.rm.iasf.cnr.it/

Aharonian, F., et al.: A Low level of extragalactic background light as revealed by gamma-rays from blazars. Nature **440**, 1018–1021 (2006). https://doi.org/10.1038/nature04680, arXiv: astro-ph/0508073

Aharonian, F., et al.: An exceptional very high energy gamma-ray flare of PKS 2155-304. Astrophys. J. **664**(2), L71–L74 (2007). https://doi.org/10.1086/520635

Aharonian, F., et al.: Observations of the crab nebula with H.E.S.S. Astron. Astrophys. **457**, 899–915 (2006). https://doi.org/10.1051/0004-6361:20065351

Aharonian, F., et al.: Spectrum and variability of the Galactic center VHE γ-ray source HESS J1745–290. Astron. Astrophys. **503**(3), 817–825 (2009). ISSN: 1432-0746. https://doi.org/10.1051/0004-6361/200811569

Aharonian, F., et al.: H.E.S.S. follow-up observations of GRB 221009A. Astrophys. J. Lett. **946**(1), L27 (2023). https://doi.org/10.3847/2041-8213/acc405

Aharonian, F.: Primary particle acceleration above 100 TeV in the shell-type Supernova Remnant RX J1713.7-3946 with deep H.E.S.S. observations. Astron. Astrophys. **464**, 235–243 (2007). https://doi.org/10.1051/0004-6361:20066381

Armillotta, L., et al.: The life cycle of the central molecular zone – I. Inflow, star formation, and winds. Mon. Not. R. Astron. Soc. **490**(3), 4401–4418 (2019). ISSN: 1365-2966. https://doi.org/10.1093/mnras/stz2880

Axford, W.I., Leer, E., Skadron, G.: The acceleration of cosmic rays by shock waves. In: International Cosmic Ray Conference, vol. 11, p. 132 (1977)

Baganoff, F.K., et al.: ChandraX-ray spectroscopic imaging of sagittarius A* and the central parsec of the galaxy. Astrophys. J. **591**(2), 891–915 (2003). https://doi.org/10.1086/375145

Bally, J., et al.: The bolocam galactic plane survey: $\lambda = 1.1$ and 0.35 mm dust continuum emission in the galactic center region **721**(1), 137–163 (2010)https://doi.org/10.1088/0004-637X/721/1/137

Bednarek, W., Bartosikm M.: Gamma-rays from the pulsar wind nebulae. Astron. Astrophys. **405**, 689–702 (2003). https://doi.org/10.1051/0004-6361:20030593

Belikov, A.V., Silk, J.: Superexponential cutoff as a probe of annihilating dark matter. Phys. Rev. Lett. **111**, 071302 (7 Aug 2013). https://doi.org/10.1103/PhysRevLett.111.071302

Belikov, A.V., Zaharijas, G., Silk, J.: Study of the gamma-ray spectrum from the Galactic Center in view of multi-TeV dark matter candidates. Phys. Rev. D **86**, 083516 (8 Oct 2012). https://doi.org/10.1103/PhysRevD.86.083516

Bell, A.R.: The acceleration of cosmic rays in shock fronts - I. Mon. Not. R. Astron. Soc. **182**, 147–156 (1978). https://doi.org/10.1093/mnras/182.2.147. Jan

Bertone, G., Merritt, D.: Time-dependent models for dark matter at the galactic center. Phys. Rev. D **72**, 103502 (10 Nov 2005). https://doi.org/10.1103/PhysRevD.72.103502

Blanch, O., Lopez, J., Martinez, M.: Testing the effective scale of quantum gravity with the next generation of gamma ray telescopes. Astropart. Phys. **19**(2), 245–252 (2003). ISSN: 0927-6505. https://doi.org/10.1016/S0927-6505(02)00205-0, https://www.sciencedirect.com/science/article/pii/S0927650502002050

Blanch, O., Martinez, M.: Exploring the gamma ray horizon with the next generation of gamma ray telescopes. Part 1: theoretical predictions. Astropart. Phys. **23**(6), 588–597 (2005). ISSN: 0927-6505. https://doi.org/10.1016/j.astropartphys.2005.03.008, https://www.sciencedirect.com/science/article/pii/S0927650505000642

Blandford, R., Simeon, P., Yuan, Y.: Cosmic ray origins: an introduction. Nucl. Phys. B Proc. Suppl. 256–257 (2014) (Tibolla, O., et al., eds., pp. 9–22). https://doi.org/10.1016/j.nuclphysbps.2014. 10.002

Blandford, R., Meier, D., Readhead, A.: Relativistic jets from active galactic nuclei. Ann. Rev. Astron. Astrophys. **57**, 467–509 (2019). https://doi.org/10.1146/annurev-astro-081817-051948

Blumenthal, G.R., Gould, R.J.: Bremsstrahlung, synchrotron radiation, and compton scattering of high-energy electrons traversing dilute gases. Rev. Mod. Phys. **42**, 237–270 (2 Apr 1970). https://doi.org/10.1103/RevModPhys.42.237. https://link.aps.org/doi/10.1103/RevModPhys.42.237

Boyce, H., et al.: Simultaneous X-ray and infrared observations of sagittarius A*'s variability. Astrophys. J. **871**(2), 161 (2019). ISSN: 1538-4357. https://doi.org/10.3847/1538-4357/aaf71f

Bringmann, T., et al.: Relevance of sharp gamma-ray features for indirect dark matter searches. Phys. Rev. D **84**, 103525 (10 Nov 2011). https://doi.org/10.1103/PhysRevD.84.103525

Bustamante, M., et al.: High energy cosmic-ray acceleration (2010). https://cds.cern.ch/record/1249755/files/p533.pdf

Cembranos, J.A.R., Gammaldi, V., Maroto, A.L.: Possible dark matter origin of the gamma ray emission from the galactic center observed by HESS. Phys. Rev. D **86**, 103506 (10 Nov 2012). https://doi.org/10.1103/PhysRevD.86.103506

Chandrasekhar, S.: XLVIII. The density of white dwarf stars. J. Astrophys. Astron. **15**, 105–109 (1931)

H.E.S.S. Collaboration, et al.: First ground-based measurement of sub-20 GeV to 100 GeV s from the Vela pulsar with H.E.S.S. II. A&A **620**, A66 (2018). https://doi.org/10.1051/0004-6361/201732153

H.E.S.S. Collaboration: H.E.S.S. (2002). https://www.mpi-hd.mpg.de/hfm/HESS/pages/about/

Di Mauro, M.: Characteristics of the galactic center excess measured with 11 years of Fermi - LAT data. Phys. Rev. D **103**(6) (2021). ISSN: 2470-0029. https://doi.org/10.1103/physrevd.103. 063029

Di Sciascio, G.: The LHAASO experiment: from gamma-ray astronomy to cosmic rays. Nucl. Part. Phys. Proc. 279–281 (2016) (Cataldi, G., De Mitri, I., Martello, D. (eds.), pp. 166–173). https://doi.org/10.1016/j.nuclphysbps.2016.10.024

Djannati-Ataı, A., et al.: Probing Vela pulsar down to 20 GeV with H.E.S.S. II observations. In: 6th International Symposium on High Energy Gamma-Ray Astronomy. American Institute of Physics Conference Series, vol. 1792, p. 040028 (2017). https://doi.org/10.1063/1.4968932

FACT.: HFACT - The first G-APD Cherenkov telescope (2003). https://www.isdc.unige.ch/fact/

Fermi LAT studies the EBL.: https://fermi.gsfc.nasa.gov/science/eteu/ebl/. Accessed 29 Feb 2024

Fermi, E.: On the origin of the cosmic radiation. Phys. Rev. **75**, 1169–1174 (8 Apr 1949). https://doi.org/10.1103/PhysRev.75.1169. https://doi.org/10.1103/PhysRev.75.1169

Fermi-LAT.: Fermi gamma-ray space telescope (2007). https://fgst.slac.stanford.edu/WhatIsLAT. asp

Ferrière, K.: Interstellar magnetic fields in the Galactic center region. AA **505**(3), 1183–1198 (2009). https://doi.org/10.1051/0004-6361/200912617

Gabici, S.: Gamma-ray emission from supernova remnants and surrounding molecular clouds. AIP Conf. Proc. **1792**(1), 020002 (2017). https://doi.org/10.1063/1.4968887

Gould, R.J., Schréder, G.P.: Opacity of the universe to high-energy photons. Phys. Rev. **155**(5), 1408–1411 (1967). https://doi.org/10.1103/PhysRev.155.1408

Hatchfield, H.P., et al.: CMZoom. II. catalog of compact submillimeter dust continuum sources in the milky way's central molecular zone **251**(1), 14 (2020). https://doi.org/10.3847/1538-4365/abb610

HAWC.: HAWC - the high-altitude water cherenkov gamma-ray observatory (2022). https://www.hawc-observatory.org/

Heger, A., et al.: How massive single stars end their life. Astrophys. J. **591**(1), 288–300 (2003). https://doi.org/10.1086/375341

Heitler, W.: The Quantum Theory of Radiation. Courier Corporation (1984)

Herold, L., Malyshev, D.: Hard and bright gamma-ray emission at the base of the Fermi bubbles. Astron. Astrophys. **625**, A110 (2019). https://doi.org/10.1051/0004-6361/201834670

Hess, V.F.: Über Beobachtungen der durchdringenden Strahlung bei sieben Freiballonfahrten. Phys. Z. **13**, 1084–1091 (1912)

Heywood, I., et al.: The 1.28 GHz MeerKAT galactic center mosaic. Astrophys. J. **925**(2), 165 (2022). ISSN: 1538-4357. https://doi.org/10.3847/1538-4357/ac449a

Hinton, J., Ruiz-Velasco, E.: Multi-messenger astronomy with very-high-energy gamma-ray observations. J. Phys.: Conf. Ser. **1468**(1), 012096 (2020). https://doi.org/10.1088/1742-6596/1468/1/012096

Ibarra, A., et al.: Gamma-ray triangles: a possible signature of asymmetric dark matter in indirect searches. Phys. Rev. D **94**, 103003 (10 Nov 2016). https://doi.org/10.1103/PhysRevD.94.103003

Ibarra, A., Gehler, S.L., Pato, M.: Dark matter constraints from box-shaped gamma-ray features. J. Cosmol. Astropart. Phys. **2012**(07), 043 (2012). https://doi.org/10.1088/1475-7516/2012/07/043

Jeff, S.: The bremsstrahlung, synchrotron and compton effects as emission processes in astrophysics (2005). http://www.jeffstanger.net/Astronomy/emissionprocesses.html

Kachelriess, M.: Lecture notes on high energy cosmic rays (2008). arXiv: 0801.4376 [astro-ph]

Kelner, S.R., Aharonian, F.A., Bugayov, V.V.: Energy spectra of gamma-rays, electrons and neutrinos produced at proton-proton interactions in the very high energy regime. Phys. Rev. D **74**, 034018 (2006). [Erratum: Phys. Rev. D 79, 039901 (2009)]. https://doi.org/10.1103/PhysRevD.74.034018, arXiv: astro-ph/0606058

Lang, K.R.: NASA's Cosmos—the material between the stars (2010). https://ase.tufts.edu/cosmos/view_images.asp?id=50

Lefranc, V.: Recherche de matière noire, observation du centre galactique avec H.E.S.S. et modernisation des caméras de H.E.S.S. I. Theses. Université Paris-Saclay (2016). https://tel.archives-ouvertes.fr/tel-01374541

Liu, S., et al.: Stochastic acceleration in the galactic center HESS source. Astrophys. J. **647**(2), 1099–1105 (2006). https://doi.org/10.1086/505171

Liu, H.: Study of longitudinal development of cosmic-ray induced air showers with LHAASO-WFCTA. In: Proceedings of 37th International Cosmic Ray Conference—PoS(ICRC2021), vol. 395, p. 245 (2021). https://doi.org/10.22323/1.395.0245

Longair, M.S. (ed.): High-Energy Astrophysics: Particles, Photons and Their Detection, vol. 1 (1992)

Longair, M.S.: High Energy Astrophysics: Stars, the Galaxy and the Interstellar Medium, vol. 2 (1994)

Macias, O., et al.: Cherenkov telescope array sensitivity to the putative millisecond pulsar population responsible for the galactic centre excess. Mon. Not. Roy. Astron. Soc. **506**(2), 1741–1760 (2021). https://doi.org/10.1093/mnras/stab1450, arXiv: 2102.05648 [astro-ph.HE]

MAGIC.: First time detection of a GRB at sub-TeV energies; MAGIC detects the GRB 190114C (2019). https://astronomerstelegram.org/?read=12390

MAGIC.: MAGIC - Major atmospheric gamma-ray imaging Cherenkov telescope (2003). http://www.magic.iac.es/

Maraschi, L., Ghisellini, G., Celotti, A.: On the broad band energy distribution of blazars. In: Courvoisier, T., Blecha A. (eds.) Multi-Wavelength Continuum Emission of AGN, vol. 159, pp. 221–232 (1994)

Millikan, R.A., Cameron, G.H.: The origin of the cosmic rays. Phys. Rev. **32**, 533–557 (4 Oct 1928). https://doi.org/10.1103/PhysRev.32.533, https://doi.org/10.1103/PhysRev.32.533

Morris, M., Serabyn, E.: The galactic center environment. Annu. Rev. Astron. Astrophys. **34**, 645–702 (1996). https://doi.org/10.1146/annurev.astro.34.1.645

Narayana Bhat, P., et al.: The third fermi GBM gamma-ray burst catalog: the first six years. Astrophys. J. Suppl. **223**(2), 28 (2016). https://doi.org/10.3847/0067-0049/223/2/28

Navarro, J.F., Frenk, C.S., White, S.D.M.: A universal density profile from hierarchical clustering. Astrophys. J. **490**, 493–508 (1997). https://doi.org/10.1086/304888

Pacini, D.: La radiazione penetrante alla superficie ed in seno alle acque. Il Nuovo Cimento **3**(1), 93–100 (1912). https://doi.org/10.1007/BF02957440

Predehl, P., et al.: Detection of large-scale X-ray bubbles in the Milky Way halo. Nature **588**(7837), 227–231 (2020). ISSN: 1476-4687. https://doi.org/10.1038/s41586-020-2979-0

Rybicki, G.B., Lightman, A.P.: Radiative Processes in Astrophysics (1979)

Rybicki, G.B., Lightman, A.P.: Radiative Processes in Astrophysics (1986)

Shower Images.: https://www-zeuthen.desy.de/~jknapp/fs/showerimages.html. Accessed 15 Mar 2024

Stecker, F.W.: Neutral-pion gamma rays from the galaxy and the interstellar gas content. Astrophys. J. **185**, 499–504 (1973). https://doi.org/10.1086/152435. Oct

Thompson, D., et al.: Exploring the extreme universe with the fermi gamma-ray space telescope (2012). https://physicstoday.scitation.org/doi/10.1063/PT.3.1787

University of Maryland.: Rotation-Powered Pulsars (2018). https://www.astro.umd.edu/~miller/teaching/astr498/lecture18.pdf

VERITAS.: VERITAS - Very energetic radiation imaging telescope array system (2003). https://veritas.sao.arizona.edu/

Chapter 5
Imaging Atmospheric Cherenkov Telescopes Technique

Abstract An overview of the detection technique of Imaging Atmospheric Cherenkov Telescopes (IACTs) is described. A description of the atmospheric shower development and the Cherenkov light emission is given. The High Energy Stereoscopic System (H.E.S.S.), one of the currently operating IACT arrays at the time of the writing, is described. Some of the standard observation methods and background-measurement techniques are quickly reviewed, and the main performances of the H.E.S.S. instrument are presented. Quick estimates of the required time to observe a dark matter signal with H.E.S.S. and a short introductory description of expectations from the next-generation observatory, the Cherenkov Telescope Array, are presented.

Keywords Atmospheric showers · Cherenkov light · High energy stereoscopic system · Instrument response functions · Event selection and reconstruction · Cherenkov telescope array

5.1 Atmospheric Showers and Cherenkov Light Emission

5.1.1 Particle Shower Development

As a primary particle collides with the Earth's atmosphere, it initiates the development of a secondary particle shower. The specific characteristics of these particle showers can vary depending on the nature of the primary particle and the specific interactions it encounters within the atmosphere. This diversity leads to distinct features in the generated particle showers, as outlined in previous studies (Bernlohr 2008).

The **electromagnetic shower** that occurs in the atmosphere involves a sequence of particles, including photons, electrons, and positrons. This shower is initiated by either a photon or an electron/positron. When a gamma ray interacts with matter, it creates an electron-positron pair, followed by the emission of radiation from these particles. The electron contributes to this radiation by producing additional gamma rays through Bremsstrahlung interactions near atomic nuclei. These gamma rays, if they still have an energy $E > 2m_e c^2$, again produce electron-positron pairs. This process repeats, producing a cascade of positrons, electrons, and photons. If the primary particle has a sufficiently high energy, these processes continue. In the case of a photon as the primary particle, pair production can begin, while for an initial cosmic ray (CR) electron, energy can be dissipated through Bremsstrahlung interactions. The progression of the shower ceases when the photons no longer possess enough energy to generate additional pairs or when other energy loss mechanisms, such as ionization, become significant. This typically occurs after reaching the energy threshold of $E_{thr} = 80\,\mathrm{MeV}/(Z+1)$, when Bremsstrahlung is no longer the dominant process of energy loss, but the ionization of air molecules is. The depth or length of the shower can be characterized by the radiation length X_0, which is material-specific. For a photon, this length corresponds to the distance over which 7/9 of the photon's initial energy ($E_{\gamma,0}$) is depleted. In the case of an electron, it represents the distance over which the electron loses all but $1/e$ of its energy. The photons travel slightly deeper in the atmosphere. An approximate definition of the depth is

$$X = X_0 \frac{\ln(E_{\gamma,0}/E_{thr})}{\ln 2}. \tag{5.1}$$

The spread width of the shower develops based on the degree of multiple scattering experienced by the electrons within it. The majority of the shower remains confined within a region extending up to twice the Molière radius, denoted as R_M. This radius signifies the diameter of a cylinder that, on average, encompasses 90% of the total energy within the shower generated by the incoming particle in the Earth's atmosphere. It is calculated as $R_M = 0.0265\,X_0(Z+1.2)$ and is a distinctive property of the material involved. A sketch and a simulation of an electromagnetic shower are shown in the first and third panels from the left in Fig. 5.1, respectively.

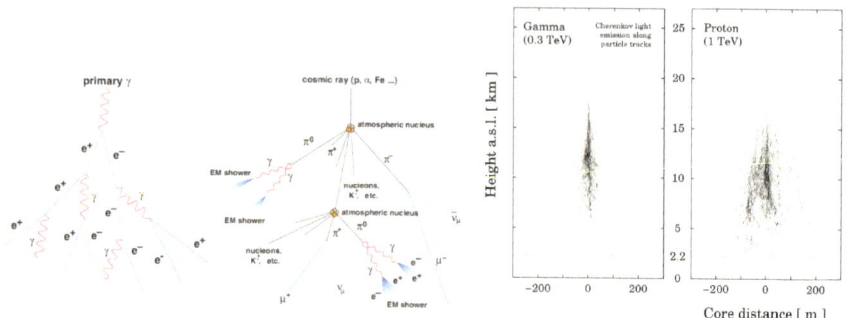

Fig. 5.1 Sketch of an electromagnetic (first) and hadronic showers (second) in the left panels. The two left panels are extracted from Pecimotika (2018). The two right panels show the simulations of an electromagnetic shower initiated by a 300 GeV gamma-ray (first) and of a hadronic shower initiated by a 1 TeV CR proton (second) in the atmosphere. The interaction of the proton is deeper and the produced shower is wider, with sub-showers displaced from the core of the shower. Bottom panels are extracted from Bernlohr (2008)

The development of an **hadronic shower** in the atmosphere is notably more intricate, primarily due to nuclear interactions and subsequent decay processes. These showers encompass various components: *(i)* hadronic subshowers originating from nuclear fragments, *(ii)* nuclear interactions of CR hadrons with the atmosphere, leading to the production of kaons and charged pions, which may subsequently decay into muons and their corresponding neutrinos, and *(iii)* muons from these interactions can further decay into electrons, generating photons and potentially additional sub-showers. These diverse interactions contribute to the broader and more complex nature of hadronic showers compared to electromagnetic ones. Additionally, hadronic showers exhibit sub-showers induced by high-momentum particles created in inelastic collisions, which can be significantly displaced from the primary shower axis. The nuclear interaction length, denoted as λ, serves as a measure for defining the depth of hadronic showers. Notably, in the air, λ exceeds the radiation length X_0, indicating that hadronic showers typically initiate at greater depths within the atmosphere when contrasted with electromagnetic showers. A sketch and a simulation of an hadronic shower are shown in the first and third panels from the left in Fig. 5.1, respectively.

5.1.2 The Cherenkov Light

The Cherenkov light is generated when charged particles move through a medium at relativistic speeds. This phenomenon occurs when the particle with velocity v traverses a medium with a refractive index denoted as n such that $v > c/n$, with c being the speed of light, then Cherenkov light is produced—a radiation which is blue/ultraviolet ($\lambda \sim 300 - 600$ nm).

The Cherenkov light emission takes the form of a cone, characterized by an angle denoted as θ_c, which satisfies the following relationship:

$$\cos \theta_c = \frac{1}{n\beta} = \frac{c}{nv}. \tag{5.2}$$

The particle moves a distance of $v\,t$ in the time t in its flight direction while the light emitted at t_0 under the angle θ_c moves in the same time t a distance $c/n\,t$. The particle also emits light in the time between t_0 and t. The emitted light produces a wavefront due to interference. The maximum angle, denoted as $\theta_{c,\max}$, is determined by the condition where $\cos \theta_{c,\max} = 1/n$. When the incident particle possesses sufficient energy, such as electrons and positrons, they can attain relativistic velocities that lead to the production of Cherenkov light. The energy threshold of electrons/positrons for Cherenkov light production is $E_{\text{thr}} = m_e c^2/\sqrt{1 - n^{-2}}$. In the Earth's atmosphere, this threshold is approximately 21 MeV at an altitude of 10 km where index $n = 1.000293$, without considering any attenuation effects.

Very-high-energy (VHE) gamma rays are indirectly detected through the Cherenkov light generated in the particle shower they initiate. The spectrum of the

Cherenkov light primarily spans over wavelengths between ~300 and 600 nm, with a peak intensity at 350 nm corresponding to the optimal sensitivity of photomultipliers (PMTs). However, accounting for optical light from stars in this wavelength range is essential, as it serves as background noise. Given the Cherenkov light is emitted under an angle of about 1° in air, a VHE gamma-ray with a primary interaction depth of 10 km will illuminate an area on the ground, the light pool, of approximately 250 m in diameter when electron scattering is considered. The ground-based Cherenkov telescopes that are situated inside the pool are designed to detect the Cherenkov photons. The Cherenkov light yield is proportional to the total track length of all particles (in the ultra-relativistic limit), which is proportional to the primary energy. Recording an image of the cascade in Cherenkov light provides, therefore, a pseudo-calorimetric measurement of the shower energy. At such wavelengths, the atmosphere is not perfectly transparent, and different effects attenuate the emitted Cherenkov light. For instance, the Rayleigh diffusion on air molecules, the Mie diffusion on aerosols, or the absorption by ozone.

The Cherenkov light spectrum follows the Franck-Tamm relation: $d^2N/d\lambda dx = 2\pi\alpha \sin^2\theta_c/\lambda^2$ (Longair 1992), with α the fine structure constant. Integrating over the 300-600 nm wavelength range gives about 10 Cherenkov photons per m. For a 1 TeV gamma-ray interacting at an altitude of 10 km, the number of electromagnetic particles with energy above the Cherenkov production threshold is about 400 (Aharonian et al. 2008). Given the light pool area and that about 10% of the primary gamma ray's energy reaches the ground, one obtains 100 Cherenkov photons per meter square per TeV of primary energy. For a typical instrumental efficiency of 10% (reflectivity of mirror surfaces, quantum efficiency of photo-sensors), primary reflectors of ~100 m^2 area are required to produce images containing 100 photoelectrons for 100 GeV gamma-ray showers.

Given that electrons and positrons are supraluminal (at GeV energies, their Lorentz factor $\gamma \sim 10^3$), Cherenkov photons reach the ground within just a few ns after their initial production in the particle shower. Therefore, it is assumed that the Cherenkov light flash at the ground lasts $\mathcal{O}(1)$ nanosecond. To detect a flash, fast photo-sensors and GHz electronics are required to resolve the faint Cherenkov signal against the night sky background (NSB) light whose typical rate is 1 photon $-$ electron $\times (\delta t/10 \text{ ns})(\theta/0.1°)^2(A/100 \text{ m}^2)$. Figure 5.2 shows the Cherenkov signal and NSB spectra as detected with the IACT technique. The Cherenkov spectrum is detected at 2200 m a.s.l. and the NSB is taken from measurements at the La Palma Observatory (Bouvier et al. 2013). The bright airglow emission lines above ~550 nm are mostly due to atomic oxygen, hydroxide, and sodium in the Earth's atmosphere. Arbitrary units were applied by the authors on the y-axis to explicit that the NSB normalization vary with the location on Earth.

The geometry of the electromagnetic shower, as well as the resulting image on the focal plane of the camera after reflection on the telescope's mirror, is illustrated in Fig. 5.3. These images provide the basis for reconstructing various parameters associated with the shower. By analyzing the Cherenkov light detected by each individual PMT within the camera, information related to the shower, such as its energy, direction, and the nature of the primary particle that initiated it, can be determined.

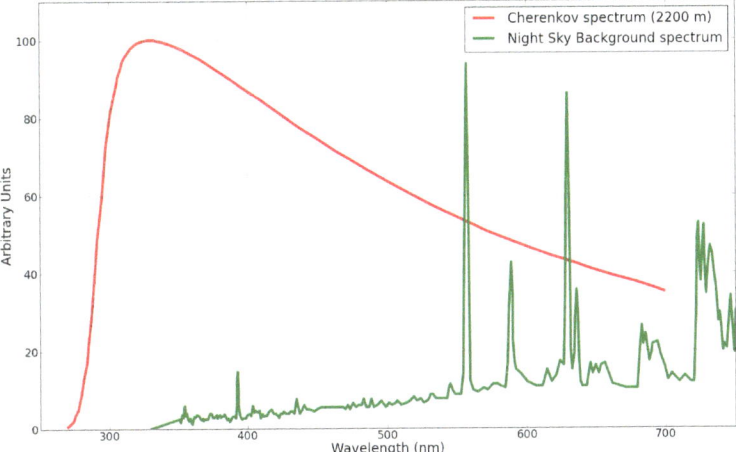

Fig. 5.2 Cherenkov signal and night sky background spectra as measured with the IACT technique, respectively, as solid red and solid green lines. Figure extracted from Bouvier et al. (2013)

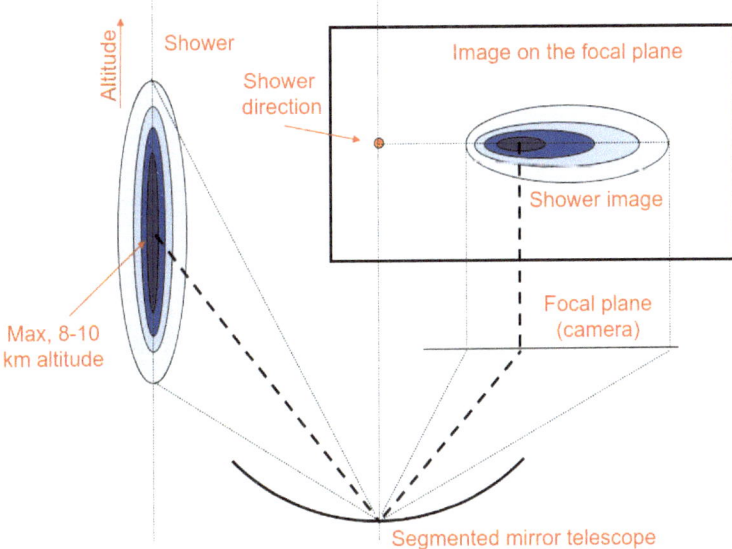

Fig. 5.3 Image of an atmospheric shower of Cherenkov light on the focal plane of the camera of an IACT. The image on the focal plane is shown after reflection on the mirror of the telescope

Through spatial and temporal analysis of these camera images, valuable information can be extracted, allowing for the characterization of the primary particle's properties and its interaction with the Earth's atmosphere.

5.2 The High Energy Stereoscopic System

5.2.1 The Instrument

The acronym H.E.S.S. stands for the High Energy Spectroscopic System, which comprises an array of five Imaging Atmospheric Cherenkov Telescopes (IACTs). The telescopes are four medium-sized ones at the corners of a square with a size of 120 meters, situated at each cardinal point. Additionally, a fifth, larger telescope is located at the center of this square arrangement. The observatory is positioned in the Namibian region of Khomas Highland, situated at geographic coordinates 23°16'17" S and 16°30'00" E, on a plateau that is approximately 1800 m above sea level. The choice of this location was influenced by factors such as the dry climate, mild temperatures, and limited light pollution. One key advantage of the H.E.S.S. array is its unique position in the Southern hemisphere, making it one of the most suitable IACTs currently operating for observing the Galactic plane, particularly the Galactic Center (GC) region, at very high energies.

Figure 5.4 shows the visibility window of the GC at the position of the supermassive black hole Sagittarius A*, from the H.E.S.S. site. The GC can be observed from

Fig. 5.4 Visibility plot for the Galactic Center region as seen from the H.E.S.S. site. The plot was produced with the tools at H.E.S.S. Collaboration 2002, for 2019. Observations at zenith angle lower than 30° are possible in the time range (x-axis, as UTC time) shown by the lilac-shaded area. With the darkening of the color, the maximum value of the zenith range for the observations increases to 45° and 60°. Yellow-shaded areas show the time range when the moon is too bright to observe the region

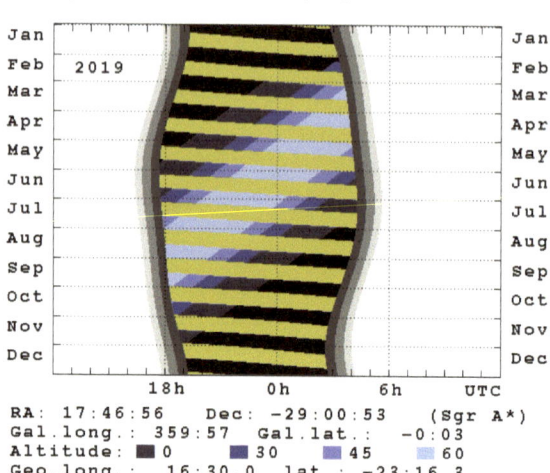

March to September with zenith angles[1] lower than 30° and under dark time conditions, resulting potentially in ~320 hours of observations per year. Note, however, that this estimate is a very rough and too optimistic approximation. First, the 17-hour RA band is populated by a number of interesting objects for VHE gamma-ray astrophysics. Therefore, not all the available time can be dedicated to GC observations. Then, bad weather conditions can hamper observations. Moreover, technical problems always affect the instrument and may prevent from normal data-taking. Obtaining 100-150 h of GC observations per year is a realistic and significant achievement in terms of data taking.

The initial phase of the experiment commenced in 2003 (Hofmann 2001), featuring the use of the four medium telescopes, often referred to as CT1-4. This phase is usually referred to as phase I. Each of these telescopes is constructed on a metallic rotating framework that offers mobility in both azimuth and zenith directions, utilizing an Alt-Az mount configuration. This configuration allows the telescopes to move both horizontally (azimuth) and vertically (zenith) to track celestial objects. The structure supports a camera and a Davies-Cotton mirror of 12-m diameter (Le Blanc et al. 2017).

In each of the small telescopes, the mirror is composed of 382 smaller circular mirrors, with a combined area totaling 108 m^2 (Bernlöhr et al. 2003; Cornils et al. 2003). During construction, all the mirror facets share the same focal length, which results in a non-continuous mirror surface. The focal length for these mirrors is set at 15 m, and the focal ratio (focal length divided by mirror diameter) is 1.2. Due to the specific alignment of the mirror tiles, the focal point is coincident with the camera. In the wavelength range of Cherenkov light, these mirrors exhibit a reflectivity of over 80%. The rapid movement of the mirror in both altitude and azimuth is controlled by a fast drive system within the mount of the telescope. This drive system is managed by servo-controlled AC motors, with backup support from battery-driven DC motors. When repositioning the telescope, the system can reach speeds of up to 100° per minute. It maintains stability within a range of 0.15 mrad rms across the entire altitude range. This stability is achieved thanks to the mirror's support structure.

Each telescope is equipped with a camera positioned at its focal point. Within these cameras, there are 960 PMTs with an individual FoV spanning 0.16° (equivalent to 3 mrad). Every PMT is essentially designated as a pixel. These pixels collectively cover a total FoV measuring 5° in diameter. Winston cones are affixed to the front of each PMT to improve performance. These cones serve multiple purposes, including reducing dead zones, enhancing the light collection surface, and directing the light onto the active area of the PMT. The pixels are organized into groups of 16, and these groups are further organized into 60 drawers. The electronics integrated into the camera housing execute all the necessary triggering and event readout operations. For an individual telescope, the typical trigger rate ranges from 800 to 1300 Hz, and the effective pixel coincidence window is approximately 1.5 ns.

---

[1] The zenith angle θ_z is related to the altitude Alt such as $\theta_z = 90° - Alt$.

Fig. 5.5 The H.E.S.S. array of IACTs. The medium telescopes CT1-4 are visible at the corners of the array. The large telescope CT5 is at the center. Figure extracted from H.E.S.S. Collaboration (2002)

In 2014, the second phase of data collection for the experiment commenced with the addition of CT5, which was fully constructed by 2012 (Krayzel et al. 2013). CT5 boasts a substantial diameter of 28 m, providing a total mirror area of $614\,m^2$. Unlike the smaller telescopes, the mirrors for CT5 form a complete parabolic shape, made up of 875 hexagonal mirror facets. The focal length for CT5's mirrors is set at 36 meters. CT5, located at the center of the array as depicted in Fig. 5.5, features a drive system that can achieve a peak positioning speed of 200° per minute in azimuth, and 100° per minute in elevation. This system offers an impressive displacement accuracy of approximately 1 mm. The camera for CT5 is similarly comprised of hexagonal pixels, totaling 128 pieces, and incorporates 2048 PMTs equipped with Winston cones. The camera has a 2-m diameter and covers a FoV of 3.2° in the sky. The integration time for the effective signal capture is 16 ns. Notably, the typical trigger rate for a monoscopic observation from CT5 is approximately 2.0 kHz.

In 2015-2016, the electronics of the cameras for the small telescopes underwent a significant upgrade aimed at enhancing the overall performance of the array. This upgrade had several beneficial effects, including the reduction of dead time in stereo mode, a decrease in the system's failure rate caused by aging, and an overall improvement in system performance (Giavitto et al. 2015). The foundation of this electronics upgrade was the implementation of NECTAR readout chips (Naumann et al. 2012). These chips were responsible for reducing the readout time from 450 s to 15 s, which allowed the telescopes to function effectively in stereoscopic mode, especially in conjunction with CT5 at a higher trigger rate. Furthermore, a comprehensive renovation of the cabling scheme, power supply, and pneumatics was carried out to ensure the system operated optimally. Figure 5.6 shows a H.E.S.S. phase I camera placed at the focal plane of a 12-m diameter telescope equipped with 960 PMTs.

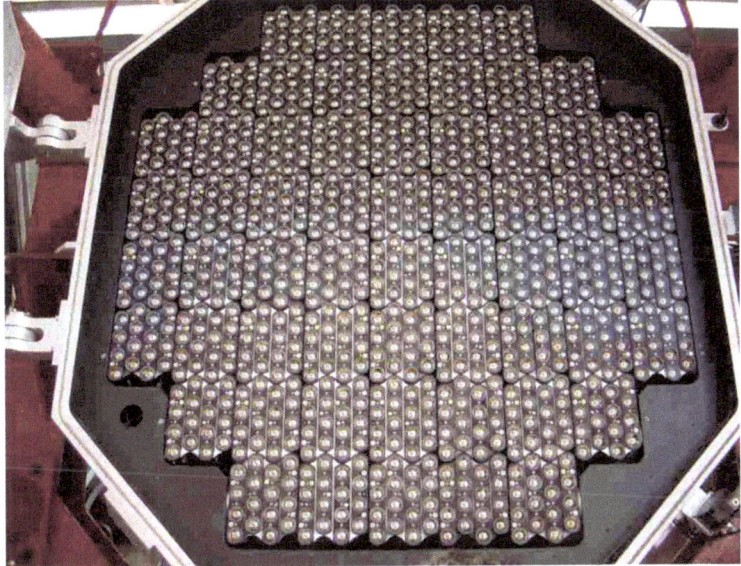

Fig. 5.6 A camera from H.E.S.S. phase I at the telescope focal plane equipped by 960 PMT tubes with 0.16° angular extent. Very compact electronics is located just behind the PMTs to decrease the camera read-out window and noise level. A Winston cone is mounted on each PMT to improve light collection efficiency. Figure extracted from H.E.S.S. Collaboration (2002)

5.2.2 Calibration of the Instrument, Trigger Systems, and Data Quality Cuts

The analysis comprises several steps: calibration, image parameterization, and event reconstruction, leading to determining the primary particle properties. Multiple parameters are essential for reconstructing the signal amplitude. Initially, the ratio between gains at low and high levels in the amplification channels is considered. Subsequently, the pedestal values in the two channels and the gains in both channels are considered. The final parameter considered is the coefficient of flat-field in each pixel, crucial for achieving a uniform output across the camera. Dedicated runs are conducted for instrument calibration, measuring these parameters. This step ensures the proper conversion of the photoelectron signal into ADC counts after excluding detected broken pixels. Further details on the calibration of CT1-4 upgraded cameras can be found in Lefranc (2016). Chalmé-Calvet (2015) provides similar details for the camera calibration in CT5. The instrument calibration facilitates the measurement of the NSB, arising from bright light spots or diffuse optical light, such as starlight, light from planets, and zodiacal light. When no Cherenkov light is measured, the NSB measured in the PMTs dominates the electronic noise. At large Galactic latitudes, it represents a single-photoelectron rate of approximately 40-100 MHz, while in the proximity of the Galactic center, it can reach up to 400 MHz. The NSB significantly influences the width of the pedestals and, consequently, the energy threshold (de Naurois and Rolland 2009).

Events selection is performed at the low level, allowing for the rejection of a substantial portion of the background, approximately 95%. This selection process involves the establishment of three distinct thresholds. Initially, the count of photoelectrons in an individual pixel is utilized to set the first threshold ($S1$), thereby identifying pixels that trigger while rejecting electronic noise and pedestal signals. The second threshold ($S2$) is determined based on the number of neighboring pixels triggered within the same camera sector, effectively identifying the triggered telescopes. The final threshold ($S3$) corresponds to the number of triggered telescopes and is defined as the stereoscopic threshold. During H.E.S.S. phase I, $S1$ was set at 4 photoelectrons per pixel, $S2$ was established as 3 pixels per sector, and $S3$ as 2 telescopes.

Additional quality cuts are implemented post-triggering. The exclusion of pixels in each camera—malfunctioning or turned off due to bright stars—is restricted to a maximum of 10%. The overall trigger rate must surpass 70% of the average within the observation *run*, denoted as a single observation lasting around 28 minutes. The trigger rate variation among the small telescopes is required to be less than 10%. Sky conditions are monitored using a weather station and an infrared LIDAR to identify cloud presence. This is essential because factors such as high humidity, temperature, or the existence of clouds can significantly impact the trigger rate or introduce inhomogeneities in the FoV (Devin et al. 2019).

The primary particle's identification relies on the characteristic shape of the shower. For example, a muon shower exhibits a distinctive ring-like signature on the camera. Additionally, muon showers are seldom observed in more than one telescope (Vacanti et al. 1994) since they originate from high-momentum particles within hadronic showers, situated far from the shower core. These isolated particles, requiring stereoscopic information, can be effectively rejected. The application of stereoscopy across the telescope array enhances the reconstruction of the shower's shape and direction. The shower's direction is reconstructed by determining the intersection of the directions extended from the major axes of the shower images reconstructed in each telescope.

5.2.3 Event Identification, Selection, and Reconstruction

Following calibration, the subsequent step involves reconstructing the images of the showers on the cameras. Depending on the reconstructed characteristics of the shower, events are classified as either gamma-like or hadron-like. The events are extracted from the runs that satisfy the earlier-mentioned selection criteria. Here, how one of the two main software chains used in the H.E.S.S. Collaboration for event selection and data analysis is succintly described.

The *Model++* analysis chain, as described in de Naurois and Rolland (2009), is employed for the analyses presented in this book. In this chain, the distribution of Cherenkov light on the camera from a particle shower is simulated to compare with the actual distribution of measured Cherenkov light in each pixel using a χ^2 test.

The typical number of showers for a smooth model range from ~20000 at low energies ($\simeq$10 GeV), up to a few hundred at the highest energies ($\gtrsim 50 - 100$ TeV). Several parameters are considered for the model generation: zenith angle, event impact distances from the camera center, primary energies, and interaction depth (de Naurois and Rolland 2009). For the model construction, these parameters are varied in wide ranges to capture features due to shower fluctuations and provide accurate simulations. Different primary energies have similar weights in the model construction, and this is ensured by the varying number of showers. Such a large number of simulations requires a very large database to be available for the *Model++* analysis chain (de Naurois and Rolland 2009).

The parameterization of the particle distribution, crucial for constructing a model of the electromagnetic shower, is developed with KASKADE (Kertzman and Sembroski 1994). This parameterization covers longitude, latitude, angles coordinates, depth of interaction, collection efficiency, and considers atmospheric conditions impacting atmospheric absorption. Additionally, the model accounts for NSB on a pixel-by-pixel basis, considering broken and inactive pixels (de Naurois and Rolland 2009). Various parameters are considered to estimate the distribution of Cherenkov photons in the camera:

- the longitudinal coordinates of the shower development;
- the longitudinal coordinates of the showers originated by charged particles;
- the energy of electrons/positrons initiating the shower;
- their position with respect to the pointing position of the telescope;
- the rate at which Cherenkov photons are produced;
- the spatial distribution of the latter with respect to the electrons;
- the opacity of the atmosphere.

The detector is simulated using SMASH (Guy 2003) to incorporate instrumental features such as the collection efficiency and reflectivity of mirrors, Winston cones, telescope geometry, photoelectrons-to-ADC counts conversion, response function, integration window, and local and central trigger systems. Simulations cover gamma rays, electrons, protons, and nuclei under various conditions, including different zenith angles, impact distances, and energy bins. Cherenkov light images in the camera and shower development obtained from simulations are stored using *lookup tables*. A maximum likelihood test is then computed on a pixel-by-pixel basis to compare measured and simulated showers.

The comparison of measured and simulated showers allows the tagging of gamma-like and hadron-like events. The quality of this comparison is evaluated through parameters, specifically the mean scale shower goodness (MSSG) (de Naurois and Rolland 2009). The MSSG indicates the agreement between gamma-ray shower templates and measurements in pixels, considering the electronic background and the NSB. To define the MSSG, the difference between the log-likelihood function and the Monte Carlo simulations predicted likelihood, i.e., $\ln\mathcal{L}(x_i, \mu_i)$ and $\langle\ln\mathcal{L}|_{\mu_i}\rangle$, respectively, is considered. The MSSG is written as:

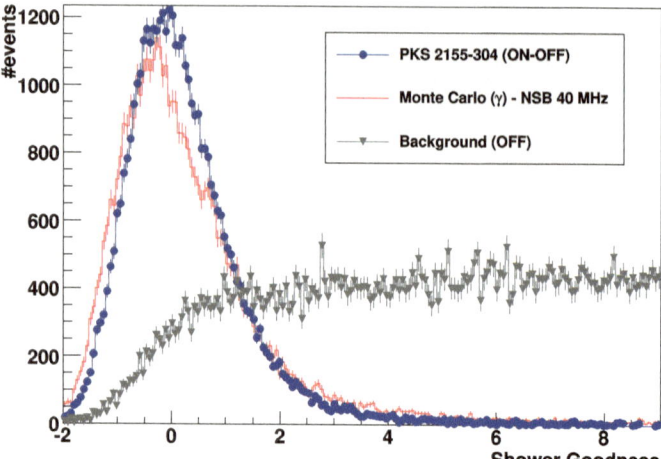

Fig. 5.7 The distribution of events versus shower goodness for the observation of the target PKS 2155-304. The measured events are represented by the blue distribution, while the simulated events are depicted in red. The background is illustrated by the gray distribution. The black vertical line shows the cut at MSSG = 0.6. Figure extracted from de Naurois and Rolland (2009)

$$\text{MSSG} = -2\frac{\sum_i [\ln\mathcal{L}(x_i, \mu_i) - \langle \ln\mathcal{L}|_{\mu_i}\rangle]}{\sqrt{2\,d.o.f.}}. \tag{5.3}$$

The number of degrees of freedom (d.o.f.) is $N_{\text{pixels}} = 6$, determined from the difference in the number of parameters used to compute the two likelihood functions (de Naurois and Rolland 2009). The index i denotes the ith pixel. Figure 5.7 illustrates how the distribution of reconstructed events and simulated events varies with the shower goodness parameter for excess photons measured towards a target. A standard cut of $MSSG = 0.6$ allows eliminating around 95% of background events while retaining 70% of photons (de Naurois and Rolland 2009). The events identified as gamma-like events among the background are termed the *residual background*. This residual background can be quantified, and various techniques for its measurement are discussed in the next section.

5.3 Observation Methods and Background Measurement

The H.E.S.S. telescopes observe approximately 1400 h per year, resulting in a duty cycle of around 15%. This duration includes observations conducted under Moonlight conditions (H.E.S.S. Collaboration 2022). Historically, the duty cycle was around 12%, before the optimizations implemented during the second phase of observations (starting from 2014). Observations with the Sun are precluded due to excessive luminosity, necessitating the Sun to be at least 18° below the horizon.

The available observing time is allocated among selected targets after evaluating proposals. When observing, a low zenith angle is generally preferred, with only a few instances involving zenith angles exceeding $60°$.

A single observation typically lasts for around 28 minutes. Various observation strategies can be employed, including pointed observations, survey observations, and transient observations. Pointed observations involve targeting a specific object, planned in advance, and considering the region's visibility above the horizon, preferably close to the zenith, to achieve the lowest energy threshold. Survey observations entail scanning a large region over several runs through a series of predefined pointing positions, and this strategy is scheduled in advance. Observations of transient phenomena, such as, gamma-ray bursts, gravitational waves, or flares from blazars, are conducted in response to alerts from other experiments, and this observational strategy cannot be pre-scheduled. The observations of transient phenomena are mainly pointed, except for the case of gravitational waves, for which a specific technique is adopted (Seglar-Arroyo and Schüssler 2017).

For pointed observations, the telescope is directed close to the selected target, and this specific position is known as the pointing position. The definition of the pointing position can vary, and the available background measurement techniques depend on the chosen pointing mode, as elaborated in the following section. The commonly employed observation mode is the *wobble* mode. In this mode, multiple pointing positions are defined around the target position to ensure comprehensive coverage. These positions are set at a designated distance known as the observational *offset*. A typical choice for the wobble mode involves establishing four perpendicular pointing positions at an offset of $0.7°$ around the target. This configuration is particularly useful for measuring the background impact on the observations, especially in searches for point-like sources.

The background measurement methodology is contingent on the observation mode. In standard pointed observations, a straightforward approach involves designating an *OFF* region to measure the background, which is then subtracted in the *ON* region where the signal is to be measured. In this setup, both the signal and residual background observations occur within the same field of view.

When employing the wobble method for observations, two modes, namely the *Wobble Ring Background* and *Wobble Multiple OFF* modes, are commonly utilized for background measurement. These will be referred to as *Ring Background* and *Multiple OFF* for conciseness. In the Ring Background mode, an annulus is defined in the observed sky, encompassing the ON region where the signal is sought. Within this region, excluding a circular mask around the target position to prevent signal contamination, the residual background is measured. If another astrophysical object appears in the FoV, it is excluded from the ring. In the Multiple OFF technique, the background is measured in circular regions of the same dimensions as the signal region, placed within the annulus where the target is being sought. In both modes, the camera's acceptance, which degrades radially from the center, is assumed to be the same for both signal and background regions when azimuthal symmetry is considered. These techniques ensure the measurement of both signal and background under identical observational conditions, occurring within the same FoV during the same observation. Both techniques are shown in Fig. 5.8. The signal region, the ON,

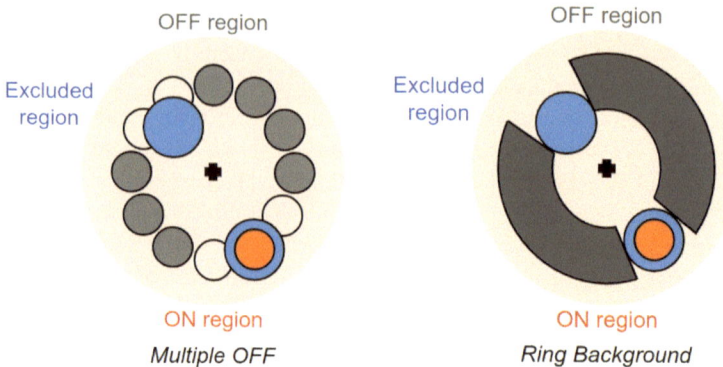

Fig. 5.8 Techniques used to measure background known as the *Wobble Multiple OFF* and *Wobble Ring Background* on the left and right, respectively. The signal region (orange) is on the target and the pointing position is shown by the black cross. The regions for the background measurement and the excluded regions are shown in gray and blue, respectively

is represented in orange, while the background OFF region is given in gray. The excluded region is shown in blue.

The analyses presented in this book adopt the approach with the definition of ON and OFF regions. This ultimately requires significantly more observation time, because part of the observed FoV has to be dedicated to the measurement of residual background. Alternative approaches make use of background models derived from Monte Carlo simulations or blank extragalactic-field observations. It still remain important to determine—at the moment of the writing—whether purely Monte Carlo simulation-based approaches can replace OFF observations. Background models are being developed with current IACT data (see, for instance, Mohrmann et al. (2019)). This kind of approach has already been used in the context of measurements with the Cherenkov Telescope Array (Acharyya et al. 2021). Nevertheless, the level of control of the systematic uncertainties required for analyses of the GC region has not been reached yet by the application of background models (Abdalla et al. 2022). H.E.S.S. is continuing to collect observations of the GC at the moment of the writing. As more observations are performed, it will be important to study the optimal scan strategies to be employed in a manner that balances both the reach for DM, but also the systematic robustness of the results.

5.4 Performances

5.4.1 Instrument Response Functions

The **effective area** of a telescope represents the portion of a plane surface, taken perpendicular to the direction of maximum radiation, through which the majority of radiation is collected. This area is energy-dependent and varies across the instrument's energy range. The variation is influenced by observation parameters such as offset and zenith angles (de Naurois and Rolland 2009) and is connected to the telescopes' optical efficiency, which is correlated with muon efficiency (Chalme-Calvet et al. 2014). Additional insights into the effective area's behavior concerning zenith angles are provided in Benbow (2005). In the left panel of Fig. 5.9 shows effective areas for H.E.S.S. considering the M++ analysis chain with three sets of cuts: standard, faint, and loose. The effective areas are extracted for simulated observations taken at zenith, with an offset from the pointing position of 0.7° (de Naurois and Rolland 2009). The figure also includes the effective area for the Hillas analysis—the other main analysis chain used in the H.E.S.S. Collaboration—with two photoelectron thresholds (p.e.): 60 p.e. and 200 p.e., with the latter being more commonly used in the literature. Above 10 TeV, the M++ effective area is smaller than the Hillas one, but it becomes comparable below this energy and superior in the hundreds of GeV energy range. The effective area is notably influenced by the radial distance from the camera's center (Berge et al. 2007), achieving a relative rate of 70% at 1.5° and maintaining negligible degradation within the inner 1° region. The right panel of Fig. 5.9 shows the effective area obtained for the Inner Galaxy Survey observations with the full five-telescope array (more details about this survey are presented in Sect. 7.1). What is shown in the panel is computed as an average of the effective area of each observation in the dataset, and all the observations were performed at different zenith angles and offsets from the pointing positions.

The **energy range** accessible to the H.E.S.S. instrument is contingent on the primary gamma-ray's energy and the dispersion of the Cherenkov shower. Initially, the primary gamma-ray must possess sufficient energy to generate Cherenkov light, and subsequently, the shower must be energetic enough to produce detectable Cherenkov

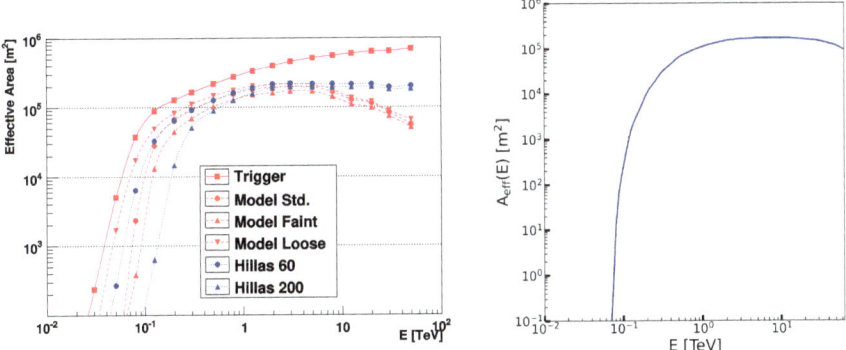

Fig. 5.9 In the left panel, the effective areas for the H.E.S.S. instrument as a function of the energy, compared for the analysis chains Model++ (red dots) and Hillas (blue dots), for different selection cuts and considering observations at zenith and with an offset from the pointing position of 0.7° are depicted. This panel is extracted from de Naurois and Rolland (2009). The right panel shows the averaged effective area obtained for the Inner Galaxy Survey observations with the full five-telescope array

light. The shower should also be sufficiently contained within the telescope's FoV, meaning it should not be too spread out. Observation conditions impact shower development, with zenith angle being a critical factor. A larger zenith angle corresponds to the shower traversing a thicker atmospheric layer. Consequently, only the most energetic showers can reach the telescopes in such conditions. The threshold is defined after applying cuts on reconstructed shower parameters. For a configuration including only CT1-4 configuration, the threshold is 160 GeV for observations at zenith. As the zenith angle increases, the threshold degrades to 220 GeV at zenith 30°, 400 GeV at zenith 45°, and 1.2 TeV at zenith 60°. The preference for observing at smaller zenith angles is primarily driven by the higher threshold at larger zenith angles unless specific observation conditions necessitate diverse angles. Nevertheless, it is stressed that a low safe energy threshold is analysis-dependent, as different systematic levels can arise at low energies. A common approach to defining a low energy threshold is to adopt the energy value at which the effective area downgrades to 10% of its maximum.

The **energy resolution** is defined as the RMS of the distribution of $\Delta E / E = |E_{\text{reco}} - E_{\text{true}}|/E_{\text{true}}$, where E_{reco} is the reconstructed energy and E_{true} is the true energy of the event (de Naurois and Rolland 2009). This metric represents the probability of recovering a mean energy E for an event with a true energy E_{true}. The energy resolution for the H.E.S.S. experiment remains approximately 10% across most of the energy range, never exceeding 15% or falling below 5%, as illustrated in Fig. 5.10. The resolution improves when more telescopes are involved in the stereo

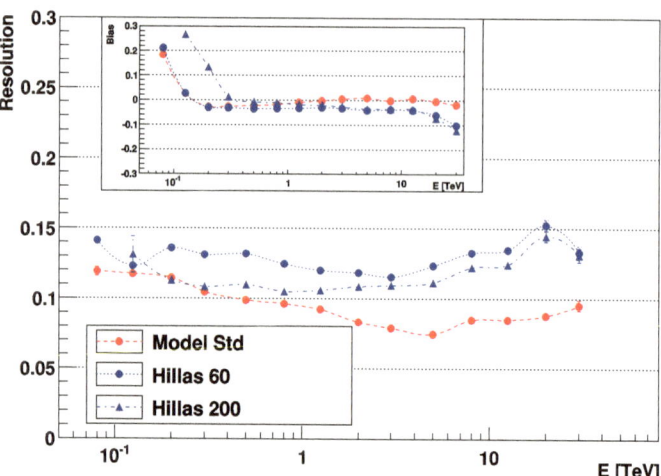

Fig. 5.10 Energy resolution and bias for the H.E.S.S. experiment are shown as a function of the energy, for the Model++ and Hillas analysis chains as red and blue dots, respectively. A few percent is reached for the energy bias. A value of 10% E is maintained for the energy resolution in Model++ throughout the whole energy range. Figure extracted from de Naurois and Rolland (2009)

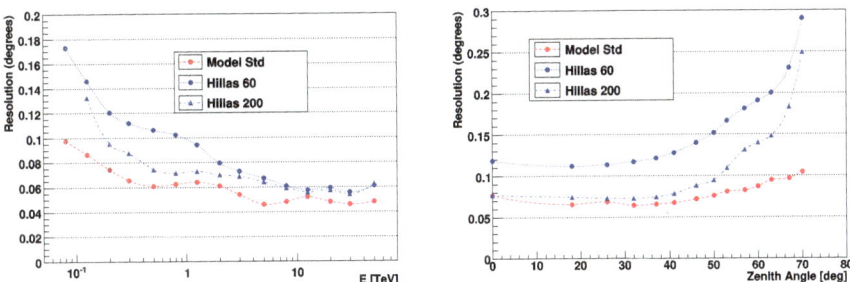

Fig. 5.11 *Left panel*: Average angular resolution for the H.E.S.S. experiment for the two analysis chains, as red dots for Model++ and as blue dots for Hillas. *Right panel*: Angular resolution as a function of energy and as a function of the zenith angle is shown in the left and right panels, respectively. Figure extracted from de Naurois and Rolland (2009)

mode and remains relatively stable with variations in offset and zenith angle. The bias of the reconstructed energy is approximately 5% across the entire sensitivity range. However, near the energy threshold, the bias increases to around 20% due to trigger effects. High resolution is crucial for discerning narrow spectral features. When considering CT5 alone, a resolution of 30% is achieved in the hundreds-of-GeV energy range.

The **angular resolution** for events reconstructed using the Model++ chain is defined as the 68% containment radius of the point spread function (de Naurois and Rolland 2009). Throughout the entire energy range, the angular resolution remains below 0.1°, exhibiting minimal dependence on the zenith angle. It stabilizes at 0.06° (at 68% confidence level) for gamma-rays in the TeV energy range. Integration of more than two telescopes in stereo mode can further enhance the angular resolution. This high angular resolution enables detailed morphological studies of extended sources and diffuse emission. As depicted in Fig. 5.11, Model++ outperforms Hillas in angular resolution. The resolution of the latter degrades notably at large zenith angles due to the reconstruction technique.

5.4.2 Reconstruction Configurations and Sensitivity

After the start of the second phase of data taking with the full five-telescopes array, H.E.S.S. data can be observed and reconstructed using three primary techniques. When exclusively CT5 is employed for reconstructing gamma-like events in a single-telescope mode, this configuration is termed the *CT5 Mono* configuration. It represents the best event reconstruction using only the 28-m diameter telescope. The reconstruction of events involving the large telescope and at least one of the small telescopes is referred to as the *CT1-5 Stereo* configuration. In this setup, at least two telescopes of the array must trigger the same shower event, and the best event reconstruction is selected between the array configuration with only the four 12-m

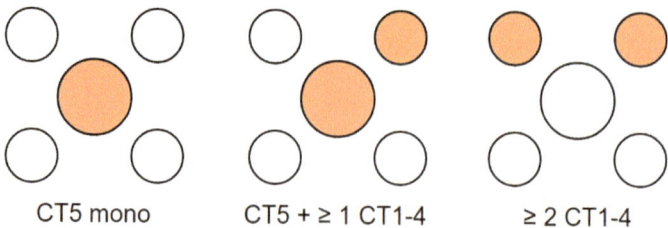

CT5 mono   CT5 + ≥ 1 CT1-4   ≥ 2 CT1-4

Fig. 5.12 The reconstruction configurations in the H.E.S.S. II phase. *CT5 Mono, CT1-5 Stereo*, and *CT1-4 Stereo* are shown respectively in the left, central and right panels of the figure

diameter telescopes and the one with all five telescopes. When the large telescope is not used, the *CT1-4 Stereo* configuration is employed for the reconstruction, selecting the best event reconstruction with the array configuration featuring only the four 12-m diameter telescopes. These three configurations are illustrated in Fig. 5.12. Additionally, there is a *Combined* configuration, where the best event reconstruction is chosen among CT1-4 Stereo, CT1-5 Stereo, and CT5 Mono.

The effectiveness of observations with the full five-telescope array varies depending on the chosen reconstruction configuration (Holler et al. 2016). Analyses that include CT5 exhibit larger acceptance below a few hundred GeV, allowing for a lower energy threshold. This is attributed to the substantial size of the large telescope, making it more sensitive to lower energies. Events below 100 GeV cannot be detected with only CT1-4 Stereo reconstruction, which has the highest energy threshold among the configurations. The Combined configuration achieves the overall best acceptance.

Evaluating the performance of an IACT can be done through its flux sensitivity. In 25 h of observations, the H.E.S.S. array reaches a sensitivity of about 1% of the Crab Nebula flux for observations taken at the zenith angles of a point-like source. Slightly larger or smaller sensitivity may be obtained for different reconstruction algorithms. The same trends observed in the effective area for different configurations are present in the flux sensitivity (Holler et al. 2016). Below 300 GeV, the CT1-5 Stereo reconstruction provides the best sensitivity. CT5 Mono and Combined analyses are also sensitive below 100 GeV. The CT1-5 Stereo configuration offers the overall best sensitivity. However, the best compromise for achieving good sensitivity across a broad range is provided by the Combined configuration. Above 3 TeV, the sensitivity is expected to degrade more rapidly for the CT5 Mono configuration.

5.5 Observing a Dark Matter Signal with H.E.S.S

Now that the theoretical ingredients to describe a photon flux from dark matter annihilation and the instrumental ingredients, such as the resolution, the effective area, and the duty cycle of an experiment like H.E.S.S., are outlined, some approximate estimates can be performed.

Following Eq. (3.6) with the case of DM annihilation into two photons $dN_\gamma/dE = 2\delta(E - m_{DM})$, and assuming the detector's effective area A and the observation time T_{obs} being independent of energy, one can determine the expected number of photons from the above as:

$$N_\gamma^{ann.}|_{E=m_{DM}} = \frac{\langle \sigma v \rangle J \, A \, T_{obs}}{4\pi m_{DM}^2} . \tag{5.4}$$

This can be obtained as a simplified version of Eq. (3.7). Note that this equation is valid only for $E = m_{DM}$, otherwise, the detector will see zero photons. Let us consider the case of a 1 TeV dark matter candidate, annihilating with the canonical thermal relic cross section $\langle \sigma v \rangle = 3 \times 10^{-26}$ cm^{3}s^{-1}. Observations on the Galactic Center (GC) region are assumed, for which a J-factor of $J \simeq 10^{21}$ GeV^{2}cm^{-5}sr inside a 1-degree region surrounding the center (see Sect. 3.2.2) is taken. At least one photon in our detector is needed to detect the signal. To have a better chance, let's require the detection of about 100 photons above the energy threshold to allow for spectra reconstruction. Then, inverting the relation in Eq. (5.4) one obtains A $T_{obs} \approx \pi \times 10^{15}$ cm^{2}s. The photons produced by this process will be at exactly 1 TeV. So they are observed through the particle showers they produce when they interact with the Earth's atmosphere and the Cherenkov flashes of light is captured by H.E.S.S. Considering an effective area of A $\sim 10^5$ m^2, which is what is obtained for observations of the GC with H.E.S.S. at energies of $\sim$1 TeV, as shown in the right panel of Fig. 5.9, H.E.S.S. would have to observe the GC for $T_{obs} \approx \pi \times 10^6$ s ≈ 870 h. This is a significant time scale, considering the H.E.S.S. duty cycle of 15%. However, roughly 2/3 of this total time have already been observed with the Inner Galaxy Survey dataset, which will be discussed in Sect. 7.1.

During that time, one can consider an irreducible residual background rate of $d\Phi_{Res.Bkg.}/dE \simeq 10^{-8}$ TeV^{-1}cm^{-2}s$^{-1}(E/1\text{TeV})^{-2.3}$, assumed to be isotropic (see Chap. 7 for more details). Therefore, more than 10^5 events are expected from the residual background alone. To distinguish a faint DM signal against such an important background, one makes use of standard statistical methods discussed in Chap. 6.

5.6 Upcoming Observatories: the Cherenkov Telescope Array

The Cherenkov Telescope Array Observatory (CTAO) will be the most powerful ground-based observatory for VHE gamma-ray astronomy in the world, putting together the expertise of H.E.S.S., MAGIC and VERITAS collaborations. It will comprise two arrays of telescopes, a southern-hemisphere array near the Paranal Observatory in Chile and a northern array on the Canary island of La Palma in Spain. It has been conceived as the first "open" gamma-ray observatory: after a proprietary period, its data and analysis software will be made available worldwide, therefore, engaging a wide research community in astronomy and high-energy physics (CTAO Exploring the Universe at the Highest Energies 2024).

Three types of telescopes will be constructed to cover the full CTAO's energy range—from $\sim$20 GeV to $\sim$300 TeV. The Large-Sized (LSTs, 23 m diameter) Telescopes, the Medium-Sized (MSTs, 12 m diameter) Telescopes and the Small-Sized (SSTs, 6 m diameter) Telescopes will cover the energy ranges from $\sim$20 GeV to $\sim$150 GeV, $\sim$150 GeV to $\sim$5 TeV, and $\sim$5 TeV to $\sim$300 TeV, respectively. The northern hemisphere array will be more limited in size and focus on the CTAO's low- and mid-energy ranges. In contrast, the southern hemisphere one, with its preferential position to observe the GC, will cover the mid- to high-energy range of the CTAO. SSTs, which are better suited to observe at the highest energies accessible by the array, will be ideal for the southern site's detection of highest-energy gamma rays. LSTs, optimized for detecting gamma rays in the hundreds-of-GeV energy range, will be installed on the northern array. The MSTs, that cover CTAO's core energy range, will be installed on both sites. The currently approved layouts of the telescope arrays—referred to as the *Alpha Configuration*—include 13 telescopes (4 LSTs, 9 MSTs) spread on a 0.5 km^2 area and 51 telescopes (14 MSTs, 37 SSTs) on a $\sim$3 km^2 area for the CTAO Northern Array and Southern Array, respectively (CTAO Exploring the Universe at the Highest Energies 2024).

The CTAO will reach a peak flux sensitivity of $\sim$10^{-13} erg cm^{-2} cm^{-1} in the $1-10$ TeV energy range. It will boast an angular resolution of around 0.15° at $\sim$100 GeV, improving until better than 0.05° for energies larger than 1 TeV. An energy resolution better than 10% will be reached for energies larger than a few hundred of GeV, getting as good as 5–7% for energies larger than a few TeV. Finally, CTA should achieve an effective area larger than 10^5 m^2 already at the low end of the energy range (about 100-200 GeV depending on the Southern or Northern configurations), improving to values of the order of 10^6 m^2 for energies larger than 1 TeV. Further details are provided in (CTAO Exploring the Universe at the Highest Energies 2024; Science with the Cherenkov Telescope Array 2018). A rendering/artistic view of the telescopes that will form the arrays of CTAO in Chile is shown in Fig. 5.13. The rendering illustrates the change in scale with CTA for ground-based arrays of IACTs.

CTA will address major questions in and beyond astrophysics grouped into three main themes: *(i)* origin and role of relativistic cosmic particles by trying to find the sites of HE particle acceleration in the Universe, understand the mechanisms of this acceleration, and find what role these particles play in feedback on star formation and galaxy evolution; *(ii)* probing extreme environments like neutron stars, black holes, relativistic jets, winds, and explosions, radiation and magnetic fields in cosmic voids. This theme is focused on understanding the physical processes, the characteristics, and the evolution at play in these environments; *(iii)* exploring frontiers in physics, addressing questions about the nature and distribution of dark matter in the Universe, quantum gravitational effects on photon propagation, and the existence of axion-like particles. The physics programme of CTA can be found at (Science with the Cherenkov Telescope Array 2018).

Fig. 5.13 Artistic view of the CTAO in Chile, at the Paranal Observatory, where all three classes of telescopes are shown. Figure extracted from Proposed CTA Telescopes (2024). Credits to Gabriel Pérez Diaz (IAC)/Marc-André Besel (CTAO)/ESO/N. Risinger

References

Abdalla, H. et al.: Search for dark matter annihilation signals in the H.E.S.S. Inner galaxy survey. Phys. Rev. Lett. **129.11**, 111101 (2022). https://doi.org/10.1103/PhysRevLett.129.111101

Acharyya, A., et al.: Sensitivity of the Cherenkov Telescope Array to a dark matter signal from the Galactic centre. J. Cosmol. Astropart. Phys. 057–057 (2021). https://doi.org/10.1088/1475-7516/2021/01/057.URL

Aharonian, F. et al.: High energy astrophysics with ground-based gamma ray detectors. Rep. Prog. Phys. **71.9**, 096901 (2008). https://doi.org/10.1088/0034-4885/71/9/096901.URL: https://dx.doi.org/10.1088/0034-4885/71/9/096901

Benbow, W.: The status and performance of H.E.S.S. AIP Conf. Proc. **745.1** (2005). Aharonian, F.A., Heinz. J.V., Horns, D. (eds.), pp. 611–616. https://doi.org/10.1063/1.1878471

Berge, D., Funk, S., Hinton, J.: Background Modelling in very-high-energy gamma-ray Astronomy. Astron. Astrophys. **466**, 1219–1229 (2007). https://doi.org/10.1051/0004-6361:20066674

Bernlöhr, K., et al.: The optical system of the HESS imaging atmospheric Cherenkov telescopes, Part 1: layout and components of the system. Astropart. Phys. **20**, 111–128 (2003). https://doi.org/10.1016/S0927-6505(03)00171-3

Bernlohr, K.: Simulation of imaging atmospheric cherenkov telescopes with CORSIKA and sim_telarray. Astropart. Phys. **30**, 149–158 (2008). https://doi.org/10.1016/j.astropartphys.2008.07.009

Bouvier, A. et al.: Photosensor characterization for the Cherenkov Telescope Array: silicon photomultiplier versus multi-anode photomultiplier tube. In: Fiederle, M. et al. ed., Hard X-Ray, Gamma-Ray, and Neutron Detector Physics XV, vol. 8852. International Society for Optics and Photonics. SPIE, 2013, 88520K. https://doi.org/10.1117/12.2023778.URL

Chalme-Calvet, R., de Naurois, M., Tavernet, J.-P.: Muon efficiency of the H.E.S.S. telescope, Mar. 2014. arXiv:1403.4550 [astro-ph.IM]

Chalmé-Calvet, R.: Étalonnage du cinquième télescope de l'expérience H.E.S.S. et observation du Centre Galactique au delà de 30 GeV. Theses. Université Pierre et Marie Curie—Paris VI, Nov. 2015. URL: https://tel.archives-ouvertes.fr/tel-01307151

H.E.S.S. Collaboration. Total observation time during the H.E.S.S. II phase. Internal H.E.S.S. Collaboration web pages (2022)

H.E.S.S. Collaboration. H.E.S.S. (2002). URL: https://www.mpi-hd.mpg.de/hfm/HESS/pages/about/

Cornils, R., et al.: The optical system of the HESS imaging atmospheric Cherenkov telescopes, Part 2: mirror alignment and point spread function. Astropart. Phys. **20**, 129–143 (2003). https://doi.org/10.1016/S0927-6505(03)00172-5

CTAO Exploring the Universe at the Highest Energies. https://www.cta-observatory.org/. Accessed 26 Feb. 2024

de Naurois, M., Rolland, L.: A high performance likelihood reconstruction of γ-rays for imaging atmospheric Cherenkov telescopes. Astropart. Phys. **32**, 231–252 (2009). https://doi.org/10.1016/j.astropartphys.2009.09.001. arXiv: 0907.2610 [astro-ph.IM]

Devin, J. et al.: Impact of H.E.S.S. Lidar profiles on Crab Nebula data. EPJ Web Conf. **197**, 01001 (2019). https://doi.org/10.1051/epjconf/201919701001. URL

Giavitto, G. et al.: A major electronics upgrade for the H.E.S.S. Cherenkov telescopes 1–4. In: 34th International Cosmic Ray Conference, vol. ICRC2015. The Hague, Netherlands, p. 996, July 2015. https://doi.org/10.22323/1.236.0996.URL: https://hal.archives-ouvertes.fr/hal-03627388

Guy, J.: Premiers résultats de l'expérience HESS et étude du potentiel de détection de matière noire supersymétrique. Theses. Université Pierre et Marie Curie—Paris VI, May 2003. URL: https://tel.archives-ouvertes.fr/tel-00003488

Hofmann, W.: Status of the high energy stereoscopic system (H.E.S.S.) project. In: 27th International Cosmic Ray Conference, Aug. 2001

Holler, M. et al.: Photon reconstruction for H.E.S.S. Using a semi-analytical model. In: PoS ICRC2015, p. 980 (2016). https://doi.org/10.22323/1.236.0980

Kertzman, M.P., Sembroski, G.H.: Computer simulation methods for investigating the detection characteristics of TeV air Cherenkov telescopes. Nucl. Instrum. Meth. A **343**, 629–643 (1994). https://doi.org/10.1016/0168-9002(94)90247-X

Krayzel, F. et al.: Improved sensitivity of H.E.S.S.-II through the fifth telescope focus system. In: Proceedings of the 33rd International Cosmic Ray Conference (ICRC2013), Rio de Janeiro (Brazil). Rio de Janeiro, Brazil, July 2013. URL: http://hal.in2p3.fr/in2p3-00907589

Le Blanc, O. et al.: Towards final characterisation and performance of the GCT prototype telescope structure for the Cherenkov Telescope Array. In: 35th International Cosmic Ray Conference (ICRC2017), vol. 301. International Cosmic Ray Conference, 836, p. 836

Lefranc, V.: Recherche de matière noire, observation du centre galactique avec H.E.S.S. et modernisation des caméras de H.E.S.S. I". Theses. Université Paris-Saclay, May 2016. URL: https://tel.archives-ouvertes.fr/tel-01374541

Longair, M.S. ed.: High-energy astrophysics, vol. 1. Particles, photons and their detection (1992)

Mohrmann, L. et al.: Validation of open-source science tools and background model construction in γ-ray astronomy. Astron. Astrophys. **632**, A72 (2019). https://doi.org/10.1051/0004-6361/201936452

Naumann, C.L. et al.: New electronics for the Cherenkov Telescope Array (NECTAr). Nucl. Instrum. Meth. A **695** (2012). Bourgeois, P. et al., pp. 44–51. https://doi.org/10.1016/j.nima.2011.11.008

Pecimotika, M.: Transmittance Simulations for the Atmosphere with Clouds. Ph.D. thesis. Nov. 2018. https://doi.org/10.13140/RG.2.2.34140.95361/1

Proposed CTA Telescopes. https://www.eso.org/public/teles-instr/paranal-observatory/ctao/. Accessed 26 Feb 2024

Science with the Cherenkov Telescope Array. World Scientific, Feb. 2018. ISBN: 9789813270091. https://doi.org/10.1142/10986.URL

Seglar-Arroyo, M., Schüssler, F.: Gravitational wave alert follow-up strategy in the H.E.S.S. multi-messenger framework. In: 52nd Rencontres de Moriond on Very High Energy Phenomena in the Universe, pp. 175–182 (2017). arXiv: 1705.10138 [astro-ph.IM]

Vacanti, G., et al.: Muon ring images with an atmospheric Cherenkov telescope. Astropart. Phys. **2**, 1–11 (1994). https://doi.org/10.1016/0927-6505(94)90012-4

Part III
Dark Matter Indirect Detection
with Very-High-Energy Gamma-Rays

Chapter 6
Statistical Data Analysis Methods

Abstract The framework for the search for new very-high-energy emissions through the log-likelihood ratio test statistics method on the measured datasets is described. A quick overview of standard Test-Statistics (TS) based methods is given followed by a discussion on how these methods are applied to derive limits on the free parameters of the tested model for the searched emission. Nuisance parameters are introduced to account for the systematic uncertainties in these methods and their impact on the final limits are discussed. The performances expected to reconstruct injected fake signals in a measured background dataset are discussed together with the method to build TS profiles for limit computation with a combination of independent datasets.

Keywords Likelihood function · Test statistics · Poisson statistics · Limits derivation · Systematic uncertaintics · Combined datasets

6.1 Introduction

The statistical significance of a measured signal can be quantified with a p-value or its equivalent Gaussian significance. A variable following a Gaussian distribution having Z standard deviations above its mean has an upper-tail probability equal to p, *i.e.*, $Z = \Phi^{-1}(1 - p)$, with Φ^{-1} is the inverse of the cumulative distribution of the standard Gaussian.

For a signal process, the astroparticle community commonly assumes to reject the background hypothesis with a significance of at least $Z = 5$ ($p = 2.87 \times 10^{-7}$), which is taken as a suitable level to claim a discovery. For excluding a signal hypothesis, a threshold p-value of 0.05, i.e., 95% confidence level, is often used ($Z = 1.64$). Note here that the discovery is based on the p-value of the background-only hypothesis, *i.e.*, if the p-value is below a given threshold, one considers this a discovery. However, the degree of belief in a new process depends also on the plausibility of the signal hypothesis and the degree to which it can describe the data.

© The Author(s), under exclusive license to Springer Nature Switzerland AG 2024   121
A. Montanari and E. Moulin, *Searching for Dark Matter with Imaging Atmospheric Cherenkov Telescopes*, https://doi.org/10.1007/978-3-031-66470-0_6

6.2 Test-Statistics-Based Methods

To find faint emissions in a dataset of very-high-energy (VHE, $E \gtrsim 100\,\text{GeV}$) observations, one starts by formulating a model that can describe the observed data. Subsequently, one defines a methodology to either detect a signal or establish limits on the values of the free parameters of the model. These limits are derived through test statistics. In this section, the key elements of the *Log-Likelihood Ratio Test Statistics* (LLRTS) technique are outlined. This method is widely employed in the analysis of H.E.S.S. datasets for the detection of (outflows or) dark matter (DM) signals. Section 6.2.1 first defines the likelihood function which is then plugged into the test statistics. The different profiling of the test statistics are explained in Sect. 6.2.2. Finally, Sects. 6.2.3 and 6.2.4 explain respectively the TS adopted for when the dataset can binned and several datasets are combined together to obtain more stringent constraints.

6.2.1 The Likelihood Function and the Test Statistics

Once the model for the searched emission is established, the likelihood function is employed to derive constraints on the free parameters of the model. Commencing with an observed dataset $\mathcal{D}$, the probability density function of the dataset is denoted as $f(\mathcal{D}, \theta)$. Here, the set θ encompasses all parameters defining the density function. If the dataset $\mathcal{D}$ consists of observed values x_i, the probability density functions for each of these values are represented by $f(x_i, \theta)$. In this context, the set of parameters θ determines these individual functions. Assuming the homogeneous and independent distribution of observed values, i.e., $\mathcal{D} = \{x_i\}_{i=1}^n$, the likelihood function for the entire dataset is expressed as:

$$\mathcal{L}(\theta, \mathcal{D}) = \prod_{i=1}^n f(x_i | \theta). \tag{6.1}$$

And by computing the logarithm of this function, one can obtain the log-likelihood:

$$\ln \mathcal{L}(\theta, \mathcal{D}) = \sum_{i=1}^n \ln f(x_i | \theta). \tag{6.2}$$

Consequently, the likelihood function—which depends on the parameters θ—provides the probability of observing the event x_i for a model contingent on θ.

When in the process of discovering a new signal, a standard definition is to set the null hypothesis, or background-only hypothesis, $H_0(\theta_0)$, as describing only known processes which are designated as background. This is tested against the alternative hypothesis $H_1(\theta_1)$, which includes both background as well as the sought-after signal. To claim the discovery of the sought signal, the hypothesis H_1 needs to be more plausible than H_0 via comparing their likelihoods. Consequently, a test statistic TS is established to evaluate which hypothesis aligns more closely with the data. The LLRTS is defined as:

$$TS = -2\ln \lambda(\theta) = -2\ln \frac{\mathcal{L}(\theta_1, \mathcal{D})}{\mathcal{L}(\theta_0, \mathcal{D})}. \tag{6.3}$$

In the high statistics limit, the TS follows a χ^2 distribution (Wilks 1938). Consequently, assuming one free parameter in the model utilized for the hypothesis H_1, the confidence limits (C.L.) on the latter can be deduced by identifying its value corresponding to TS = 2.71, for a one-sided likelihood and one degree of freedom between the two hypotheses[1]. Analogously, TS = 3.84 corresponds to 99% C.L. limits. TS values for other confidence levels can be obtained by considering the χ^2 values for different degrees of freedom when the high statistics limit is valid.

6.2.2 Profiling Likelihood Technique

The analyses presented here utilize the *full profiling* definition of the TS, as elaborately described in Cowan et al. (2011). The ratio of the likelihood functions for the H_1 and H_0 hypotheses is then defined as:

$$\lambda(\theta) = \frac{\mathcal{L}(\widehat{\widehat{\theta}}_1, \mathcal{D})}{\mathcal{L}(\widehat{\theta}_0, \mathcal{D})}. \tag{6.4}$$

θ_1 and θ_0 are the sets of parameters for the hypotheses H_1 and H_0, respectively. The definition is applicable for $0 \leq \widehat{\widehat{\theta}} \leq \theta$. The term $\widehat{\widehat{\theta}}$ is computed from a conditional maximization of the likelihood function, hence it depends on θ. In contrast, $\widehat{\theta}$ is obtained from a non-conditional maximization of the likelihood function and does not depend on θ. In situations where $\widehat{\theta} < 0$, the *hybrid profiling* definition can be employed. In this case, the ratio of the likelihood functions is defined as:

$$\lambda(\theta) = \frac{\mathcal{L}(\widehat{\widehat{\theta}}_1, \mathcal{D})}{\mathcal{L}(\widehat{\widehat{\theta}}_0(0), \mathcal{D})}. \tag{6.5}$$

A *simplified* definition, with no profiling of the likelihood, can be adopted. In this case, the ratio of the likelihood functions is defined as:

---

[1] A one-sided test considers one rejection region, *i.e.* one checks whether the parameter of interest is larger (or smaller) than a given value. Conversely, a two-sided test is used when looking for the equivalence of a parameter to a certain value. Deviations from that value in both directions are rejected.

$$\lambda(\theta) = \frac{\mathcal{L}(\widehat{\theta_1}, \mathcal{D})}{\mathcal{L}(\widehat{\theta_0}, \mathcal{D})}. \qquad (6.6)$$

In case of *discovery*, one computes $TS = \lambda(0)$. This implies rejecting the hypothesis of background-only emission (Cowan et al. 2011). To define a discovery, one needs to take into account the significance of the excess in the observed signal. More details about how to derive the significance of the excess will be presented in Sect. 6.3.3.

6.2.3 Binned Likelihood Technique

The analyses presented in this work rely on massive datasets, making an un-binned likelihood (event-by-event) approach impractical. Consequently, the analyses are conducted with binned datasets. The likelihood function used to derive constraints on the parameters of the tested model is therefore binned. Spectral bins are defined for the energy range where the instrument is sensitive, while spatial bins are typically defined for the region of the sky where events are measured. For instance, in the analysis searching for DM annihilation signals in the H.E.S.S. observations toward the Galactic Center (GC) region, spatial and spectral bins are employed in the likelihood function to better exploit the expected DM signal's morphology. In that case, the binned likelihood function is denoted as $\mathcal{L}_{i,j}$ for the ith spatial and jth spectral bins. To obtain limits for the tested hypothesis, the total likelihood function is computed through the product of the binned functions over all the bins: $\mathcal{L} = \prod_{i,j} \mathcal{L}_{i,j}$.

6.2.4 Combination of Independent Datasets

Multiple astrophysical objects or regions of the sky can be employed for the measurement or derivation of constraints on the same searched model. Consequently, a total likelihood function is defined for each dataset, corresponding to each object, as explained in the previous section. Limits on the model of the searched emission can be obtained with each individual dataset, or the likelihood functions of the datasets can be combined into a unified likelihood function to derive combined limits using the TS. Dataset combination is typically performed when no significant overall excess is found anywhere in the FoV in any of the individual datasets or in the stacked datasets. The combined likelihood function is then expressed as $\mathcal{L}_{\text{comb}} = \prod_{k=1}^{N_{\text{targets}}} \mathcal{L}_k$, where $\mathcal{L}_k$ is the total likelihood computed for the target and the dataset k. Constraints obtained with the combined likelihood function are inherently stronger than those obtained with the functions for the individual datasets due to the increased statistics.

Two approaches can be utilized for the combination of the likelihood functions. The first involves the summation of the statistics obtained in all the individual datasets or from different instruments. The total number of measured events can be determined for each energy bin i and spatial bin j by summing the events over the k dataset. The same procedure is applied for deriving expected events from the background and the signal emission. Then the total likelihood is constructed as the product over the likelihood functions for each energy and space bin, denoted as $\mathcal{L}_{\text{comb}} = \prod_{i,j} \mathcal{L}_{i,j,\text{tot}}$. In this case, for each spatial and spectral bin, the subscript *tot* indicates that the likelihood function is obtained with the sum of the events in the bins over the dataset. However, combining $\mathcal{L}_{i,j,\text{tot}}$ results in a loss of information. When the combination includes datasets with significantly different event counts, the potential fluctuations due to varying statistics are smoothed out. This information loss can be avoided when the combination is performed at another level. The total likelihood function can be constructed as previously mentioned for each dataset k: $\mathcal{L}_k = \prod_{i,j} \mathcal{L}_{i,j,k}$. Then the combined likelihood function is obtained through the product of the functions over the index k, *i.e.*, $\mathcal{L}_{\text{comb}} = \prod_{k=1}^{N_{\text{targets}}} \mathcal{L}_k$. This second approach preserves the information from each dataset, allowing for a more nuanced combination of the likelihoods.

6.3 Statistical Framework for Data Analysis

If no significant excess against measured background—extracted in a control region hereafter referred as to the *OFF* region—is found in a spatial region where the signal is expected—hereafter referred as to the *ON* region, the LLRTS procedure can be applied to derive limits on the parameters for the model of the searched emission. To do this, energy distributions of measured and expected events have to be defined to exploit the binning of the likelihood functions previously defined. Then, the hypothesis with a model for the searched emission can be tested against the hypothesis with only background. To do this, one can define the likelihood function and the test statistic. Limits on the free parameters of the assumed model can then be derived. In the next sections, the framework is defined first in Sect. 6.3.1, then the likelihood function for Poisson statistics is discussed in Sect. 6.3.2, the computation of the significance of a measured excess is given in Sect. 6.3.3 and limits are explicitly computed in Sect. 6.4.

6.3.1 A Mock Very-High-Energy Gamma-Ray Dataset

In VHE astrophysical analyses, the search for emission is typically conducted in the ON region, which is also what is referred as to the region of interest (ROI). The background is measured in the OFF region. Methods for defining the OFF region have been discussed in Sect. 5.3. An example with real data will be provided in Chap. 7. For the example presented in this chapter, the ROI is defined following a standard approach for searching for DM signals from the region around the center of the Milky Way – the GC region. It is considered as a circular region with a $3°$ radius, which is then divided into rings with a width of 0.1°. The rings are considered from an inner radius of 0.3° up to 2.9°, centered on the GC.

After defining the ON and OFF regions, events in these regions can be measured, collected independently, and binned in energy to construct event energy distributions for the ON and OFF regions, respectively. In the example presented in this chapter, a mock dataset is generated from 100 Poisson realizations of the actual residual-background measurements obtained through observations of the GC region by the H.E.S.S. collaboration over recent years. Further details about the observed dataset and the corresponding analysis are provided later in Chap. 7. Independent realizations are created for the ON and OFF simulated distributions, utilizing the OFF events measured in each energy bin of the observed distributions. These events were collected with the observations that are described more in detail in Chap. 7 and reconstructed for the CT1-5 Stereo configuration. Therefore, the mean of the Poisson probability function is set to N_{OFF} for each energy bin. The realizations are computed indipendently for each observation in the dataset (see Chap. 7 for the dataset description). Additionally, the simulated distributions are rescaled, assuming 500 h of homogeneous observations of the inner halo of the Milky Way, covering Galactic latitudes from $b = -3°$ up to $b = 6°$ and Galactic longitudes of $1 \leq |4|°$, with the full-five telescopes H.E.S.S. array.

In this chapter, the event distributions for ROI 22 are illustrated in Fig. 6.1. The ON and OFF event distributions are depicted in black and red, respectively, along with 1σ statistical error bands for each energy bin. These distributions exhibit the anticipated power-law-like behavior typical of residual-background measurements. However, beyond approximately 10 TeV, the number of measured events remains constant. This

Fig. 6.1 The event energy distributions presented for ROI 22. The ON and OFF distributions are generated from independent Poisson realizations of the measured OFF distribution—to emulate two independent measurements in the absence of any signal

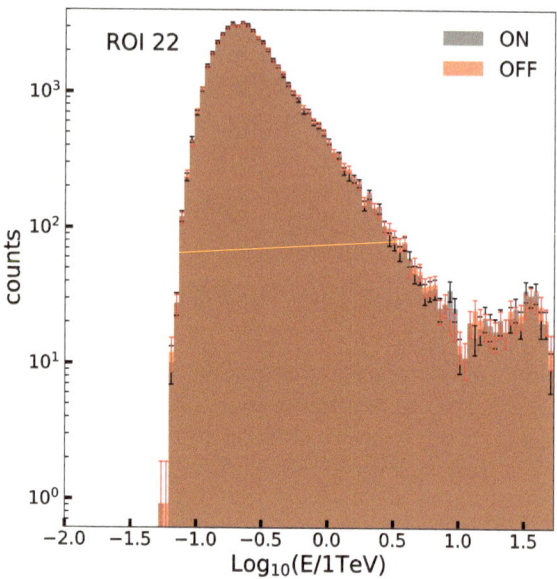

phenomenon arises from the population of events measured and reconstructed with only CT1-4: the four small telescopes of the H.E.S.S. array are more sensitive to high energies. Details regarding the reconstruction modes were presented in Sect. 5.4.2.

6.3.2 Poisson Likelihoods Function

To test emission models against background-only hypotheses in VHE gamma-ray datasets, Poisson distributions of counts x_i are employed. The measured number of photons in the ON and OFF regions is denoted as N_{ON} and N_{OFF}, respectively. For the search for DM annihilation signals, it is anticipated to measure N_S signal photons and N_B background photons from the residual background in the ON region. The value of N_S is derived from Eq. (3.7), with more details provided in Sect. 3.2. The measured photons N_{ON} follow a Poisson distribution with a mean of $N_S + N_B$. In cases where a leakage of signal photons is expected in the OFF region, N'_S signal photons and αN_B background photons should be measured, where α denotes the ratio between the solid angle size in the sky of the OFF and ON regions. Consequently, the N_{OFF} photons are distributed according to a Poisson function with a mean value of $N'_S + \alpha N_B$. With these definitions, the likelihood function can be expressed as:

$$\mathcal{L}(N_S, N_B | N_{ON}, N_{OFF}, \alpha) = \frac{(N_S + N_B)^{N_{ON}}}{N_{ON}!} e^{-(N_S + N_B)} \frac{(N'_S + \alpha N_B)^{N_{OFF}}}{N_{OFF}!} e^{-(N'_S + \alpha N_B)}.$$
(6.7)

Following the *full profiling* approach, the TS is redefined from what introduced in Sect. 6.2.2. In $TS = -2\ln(\lambda)$, one considers:

$$TS = -2\ln(\lambda(N_S)) = -2\ln\left(\frac{\mathcal{L}(N_S, \widehat{\widehat{N_B(N_S)}})}{\mathcal{L}(\widehat{N_S}, \widehat{N_B(N_S)})}\right).$$
(6.8)

The equation holds within the range $0 \leq \widehat{N_S} \leq N_S$. $\widehat{N_B}$ results from a non-conditional maximization, rendering it independent of N_S as it is maximized separately. Consequently, $\widehat{N_B} = N_{OFF}/\alpha$ and $\widehat{N_S} = N_{ON} - N_{OFF}/\alpha$. In cases where $\widehat{N_S} < 0$ (expectation can be negative due to fluctuations), the TS must be defined using the *hybrid profiling* approach. Conversely, for $\widehat{N_S} > N_S$, $TS \equiv 0$.

$\widehat{\widehat{N_B(N_S)}}$ is determined through conditional maximization, where $d\mathcal{L}/dN_B = 0$, and it represents the optimal estimate of the background for a given signal N_S. A simplified derivation, for $\alpha = 1$—which is realistic in case only one OFF region is defined with respect to the ON region—is given in Eq. (6.9).

$$\widehat{\widehat{N_B(N_S)}} = \frac{N_{ON} + N_{OFF} - 2(N_S + N'_S) \pm \sqrt{(2(N_S + N'_S) - N_{ON} - N_{OFF})^2 - 8(2N_S N'_S - N_{ON} N'_S - N_{OFF} N_S)}}{4}.$$
(6.9)

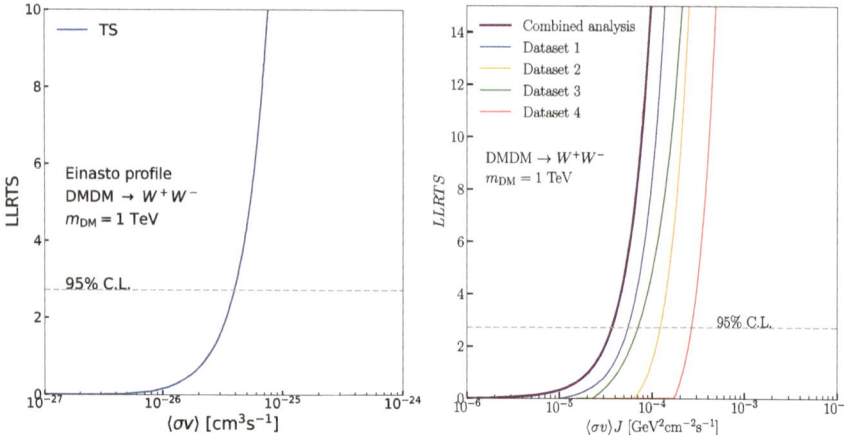

Fig. 6.2 *Left panel*: An example of a Log-likelihood ratio test statistic profile versus $\langle\sigma v\rangle$ values following Eq. (6.8). This is computed for the assumption of DM particles with mass of 1 TeV, annihilating into the W^+W^- channel and distributed according to the Einasto profile in the GC region. *Right panel*: Log-likelihood ratio test statistic profile for four independent datasets and their combination, obtained with the procedure described in Sect. 6.2.4. The profiles have been computed with datasets extracted from Abdalla et al. (2021). These profiles are computed for the assumption of DM particles with mass of 1 TeV, and annihilating into the W^+W^- channel. No assumption is made on the DM distribution profile for these extragalactic objects (Abdalla et al. 2021). $TS = 2.71$ provides a one-sided 95% C.L. upper limit on $\langle\sigma v\rangle$ and $\langle\sigma v\rangle J$, for the left and right panels respectively

$\widehat{N_S}$ is obtained as the value of $N_S(\widehat{\langle\sigma v\rangle}(m_{DM}))$, where $\widehat{\langle\sigma v\rangle}(m_{DM})$ is the estimator derived from a non-conditional maximization of $\mathcal{L}(N_S, \widehat{N_B(N_S)})$, *i.e.* the term corresponding to the denominator of Eq. (6.8) before maximization of N_S. $\widehat{\langle\sigma v\rangle}$ is obtained computationally by realizations of $\mathcal{L}(N_S, \widehat{N_B(N_S)})$, over a wide range of $\langle\sigma v\rangle$, until its maximum value is found. For an analysis computing upper limits on the annihilation cross section $\langle\sigma v\rangle$ of DM particles, this procedure is repeated for all the masses m_{DM} tested in the analysis.

6.3.3 Significance of the Measured Excess

From the measured events in the ON and OFF regions, it is possible to compute the excess in the signal region compared to the background. In accordance with Li and Ma (1983), the significance of the excess can be calculated using the equation:

$$S = sign\sqrt{2}\left\{N_{ON}\ln\left[\frac{1+\alpha}{\alpha}\left(\frac{N_{ON}}{N_{ON}+N_{OFF}}\right)\right] + N_{OFF}\ln\left[(1+\alpha)\left(\frac{N_{OFF}}{N_{ON}+N_{OFF}}\right)\right]\right\}^{1/2}$$

$$(6.10)$$

with negative *sign* in case $N_{OFF} > N_{ON}$, hence a negative significance is computed. A standard assumption for an excess in gamma-ray astrophysics to be considered as significant, is to obtain S above 5σ.

6.4 Upper Limits Computation

6.4.1 Observed Limits

From Eq. (3.7), the expected number of events from self-annihilating DM particles N_S depends on the velocity-weighted annihilation cross-section $\langle \sigma v \rangle$. $\langle \sigma v \rangle$ is treated as the free parameter while the other model parameters such as the DM mass m_{DM}, the annihilation channel, and the J-factor value for the DM distribution, are fixed. Therefore, utilizing Eq. (6.8) and the LLRTS procedure, one can compute upper limits on $\langle \sigma v \rangle$. This can be performed for each considered DM particle of mass m_{DM} that generates an annihilation spectrum dN/dE for the W^+W^- annihilation channel, and for DM distributed according to the Einasto J-factor profile (see Sect. 3.4 and 3.5 for the annihilation spectra and J-factor discussions).

Upper limits (U.L.) at a confidence level of 95% C.L. are subsequently determined as a function of m_{DM}, employing the event counts given by the terms N_{ON}, N_{OFF}, and N_B. An example of event count distributions of N_{ON} and N_{OFF} used for the limits derivation is displayed in the left panel of Fig. 6.1. The 95% C.L. U.L. for $\langle \sigma v \rangle$ corresponds to TS = 2.71. Any values of $\langle \sigma v \rangle$ with TS greater than 2.71 are excluded at 95% C.L. Assuming that one has measured distributions of events in the ON and OFF regions, *observed* limits can be computed with the measured ON and OFF distributions, and the distributions of N_S and N_B.

An example of a TS profile versus $\langle \sigma v \rangle$ values, computed with Eq. (6.8), is shown in Fig. 6.2 for a DM particle with mass of 1 TeV, annihilating into the W^+W^- channel and distributed according to the Einasto profile in the GC region. $TS = 2.71$ provides a one-sided 95% C.L. upper limit on $\langle \sigma v \rangle$. Although this profile has not been obtained with true measured events, it is representative of a standard TS($\langle \sigma v \rangle$) behavior when the desired outcome of the analysis is to set upper limits.

6.4.2 Mean Expected Limits and Containment Bands

The computation of *expected* limits can make use of independent Poisson realizations of the measured background event distributions, as introduced earlier. For each realization of independent ON and OFF distributions, the computation of 95% C.L. upper limits using the LLRTS procedure and Eq. (6.8) is performed. The mean expected limits are then determined by extracting the mean of the distribution of $\langle \sigma v \rangle$ values

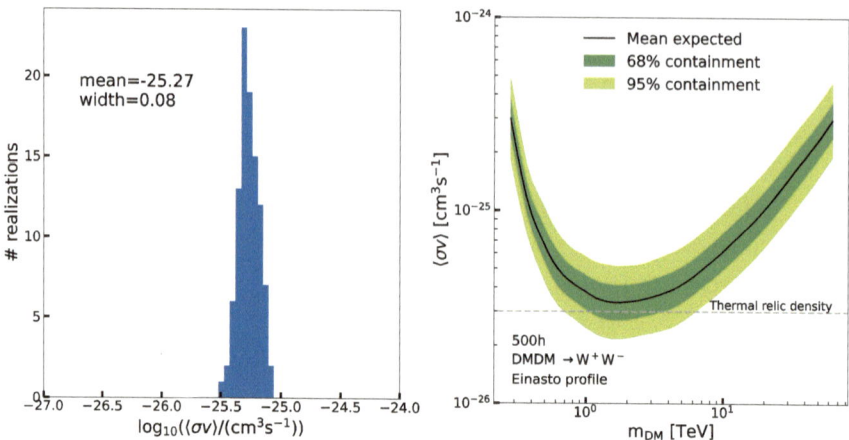

Fig. 6.3 *Left panel*: the $\log_{10}\langle\sigma v\rangle$ distribution was calculated using 100 independent realizations of the ON and OFF distributions for a 1 TeV DM particle. The distribution has a mean of -25.27 and a standard deviation of 0.08. *Right panel*: Expected limits on $\langle\sigma v\rangle$ for DM particles annihilating in the W^+W^- channel and for DM distributed in the target as an Einasto profile. The mean expected limit (black solid line) is obtained from 100 realizations of the background. 68 and 95% containment bands are shown and are obtained from the one and two standard deviations of distributions like the one shown in the left panel. The horizontal grey long-dashed line is set to the value of the natural scale expected for the thermal-relic WIMPs

obtained through the Poisson realizations of the expected background. The containment bands at 68 and 95% C.L. are determined thanks to the standard deviation of the same distribution. As an illustration, the left panel of Fig. 6.3 shows an example of the log-values of $\langle\sigma v\rangle$ obtained with 100 independent Poisson realizations of ON and OFF distributions. The mean of the distribution is -25.27, with a standard deviation of 0.08. The expected limits and containment bands, for several values of the DM particle mass m_{DM}, derived with this procedure are presented in the right panel of Fig. 6.3. The expected limits are depicted as the black solid line, while the containment bands are represented as the green and yellow shaded areas for the 68 and 95% C.L., respectively. The expected limits for each value of m_{DM} are extracted as the $\langle\sigma v\rangle$ at which $TS = 2.71$. This means that for each m_{DM}, a TS profile as shown in the left panel of Fig. 6.2 is computed. It is clear that, when scanning through the $\langle\sigma v\rangle$ values for the TS profile computation, it is unlikely to strictly get $TS = 2.71$. Therefore a level of precision has to be decided depending on the desired sensitivity of the analysis.

6.4.3 Expected Limits with the Asimov Dataset

Expected limits and containment bands can also be determined through an alternative method, known as the Asimov dataset (Cowan et al. 2011), which does not use Monte Carlo realizations of the expected backgrounds. This approach allows quickly assessing the experiment's sensitivity to DM in an annihilation channel. The Asimov dataset is an artificial dataset designed to reproduce the actual parameter values when the estimators in the LLRTS are evaluated. By setting the partial derivatives of the likelihood function with respect to the parameters to zero, the results for the estimators are obtained.

In the Asimov dataset, data counts correspond to the outcome of a Monte Carlo realization with very large statistics, aligning with the mean expectation of the corresponding actual measurements. For this setup, $N_{ON} \equiv N_{OFF}$ in Eq. (6.7), indicating no excess. Using the Asimov dataset with a likelihood function defined by a Poisson distribution $\mathcal{L}(\lambda|d) = \frac{\lambda^d}{d} \exp(-\lambda)$ involves setting d equal to the mean λ and calculating the limit accordingly. This eliminates the need for realizations on d, computing $\overline{\lambda}_{95\%}$, and taking the mean for each one. The computation of the TS in this context is given by $TS = (\Phi^{-1}(0.95) \pm N)^2$, where Φ is the cumulative distribution function of a standard Normal distribution with mean $\mu = 0$ and width $\sigma = 1$. To compute the containment bands, one adds N; $N = 1$ or 2 to provide the 1 or 2σ containment band, respectively. The conventional LLRTS for $TS = 2.71$ corresponds to $N = 0$, resulting in the mean expected limits. Note however, that the limits computed with this procedure have to be power-constrained. This does not allow the limits to move below the expected one-sigma lower limit, therefore preventing the computation of the -2σ containment band (Lisanti et al. 2018). The reader can follow the same procedure explained in Montanari et al. (2023); Lisanti et al. (2018) for power constraining the limits when needed.

A more detailed comparison between the expected limits derived with the Asimov dataset and realizations of the true measurements is provided later in Chap. 8, along with illustrative examples.

6.4.4 Limits from a Combination of Datasets

Section 6.2.4 introduced the way of computing combined limits when one wants to exploit multiple independent datasets to obtain constraints on the same searched model.

In this section, an example of combined limits derivation is shown for when the combination of datasets is performed at the likelihood level. To explain this, the results of the analysis derived in Abdalla et al. (2021) are used. The authors of this work computed upper limits on the product $\langle \sigma v \rangle J$ for four selected DM subhalos among the *Fermi*-LAT unidentified sources in the Third Catalog of High-Energy *Fermi*-LAT Sources (Ajello et al. 2017). As opposed to objects with measured stellar

Fig. 6.4 Limits on $\langle\sigma v\rangle J$ for DM particles annihilating in the W^+W^- channel. The limits are obtained for four independent datasets and their combination. The upper limits are processed with the combination procedure described in Sect. 6.2.4 and by obtaining the TS individual and combined profiles as shown in the right panel of Fig. 6.2. The results have been extracted from Abdalla et al. (2021)

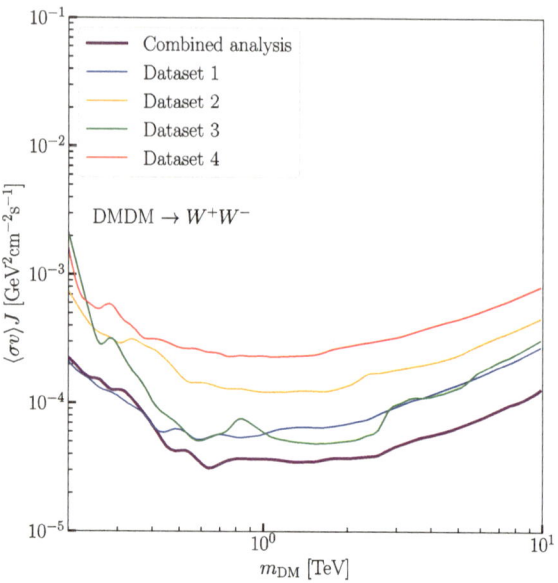

dynamics like dwarf galaxies, these DM subhalos have unknown distances to Earth, therefore their J-factors cannot be derived from stellar kinematics. Only upper limits on the overall normalization of the gamma-ray flux $\langle\sigma v\rangle$ J could be obtained in the analysis.

Only the results derived with H.E.S.S. datasets from that analysis are extracted. The authors derived upper limits from each individual object, as shown in Fig. 4 left panel of Abdalla et al. (2021), and then computed combined upper limits with the procedure for the likelihood combination that is just recalled – this is shown in the top-right panel in Fig. 3 of Abdalla et al. (2021).

The TS profiles for the four independent dataset and for the combination at the likelihood level of all of them is displayed in the right panel of Fig. 6.2. The profile derived with the combination is the thick purple line. These profiles are shown for one DM mass, for the W^+W^- annihilation channel and no assumption on the DM distribution as previously explained. Repeating the procedure for all DM masses, the 95% C.L. upper limits on $\langle\sigma v\rangle$ J are derived. These are shown in Fig. 6.4, for the individual datasets and their combination. The color code in this figure is the same as for the right panel of Fig. 6.2.

6.5 Systematic Uncertainties in the Likelihood Function

The determination of limits can be influenced by several sources of uncertainties. In the DM analysis, the GC region serves as the observational target for dataset collection. This region is densely populated with numerous VHE sources—more details have been provided in Sect. 4.4. One approach is to mask these known emissions, thereby preventing any potential signal leakage into the ROI for the analysis.

The intensity of the Night Sky Background (NSB) in the inner halo of the Milky Way can undergo substantial variations. Nevertheless, the analysis of raw data in this example employs the shower template method, outlined in de Naurois and Rolland (2009), which incorporates a specialized treatment for the NSB. Consequently, during the background measurement, no additional normalization is required. Further insights into how the NSB level can impact photon measurements are presented later in Sect. 7.2.

The gamma-like rate observed in the FoV is contingent on the zenith angle gradient during observations. It is anticipated that for each one-degree variance in zenith angle, there will be a 1% fluctuation in the gamma-like rate. A more comprehensive explanation of this estimation is furnished for the analysis presented in Chap. 7, and is delineated later in Sect. 7.2.

A potential systematic uncertainty emerges when assuming azimuthal symmetry in the field of view. To investigate this, the count numbers were calculated for a specific pointing position in the dataset. Further discussion on the test for azimuthal symmetry is presented in Sect. 7.2. No substantial impact was noted beyond the expected 1%-per-degree gradient in the FoV. This effect is anticipated due to variations in the zenith angle during observations.

The raw data analysis in H.E.S.S. can be conducted using various analysis chains, introducing potential systematic uncertainties on the energy scale of the reconstructed events. For the dataset used in Chap. 7, the systematic uncertainty on the energy scale is determined to be 10% when constructing the energy count distributions. Since this uncertainty similarly impacts the energy scale of both the measured and expected energy count distributions, it is not factored into the computation of the limits.

6.5.1 Residual Background Uncertainty

When conducting measurements in the ON and OFF regions, a discrepancy in the zenith angles of the events measured in the two regions is observed, resulting in an expected gradient in the residual background between the two regions. For the example illustrated in this chapter (which is obtained by realizations of the dataset used in Chap. 7), the difference in the means of the ON and OFF zenith angle distributions is up to 1°. For each observational run in the dataset, the measured OFF is renormalized based on the difference in the means of the ON and OFF zenith angle distributions.

Additionally, there is the typical width of the zenith angle distributions, approximately $1°$. Consequently, a systematic uncertainty of 1% for the normalization of the measured energy count distributions is employed—since for each degree of variance in the zenith angle, a 1% uncertainty is expected. This systematic uncertainty can be incorporated into the likelihood function as a Gaussian nuisance parameter, modifying the likelihood function as follows:

$$\mathcal{L}(N_S, N_B | N_{ON}, N_{OFF}, \alpha) = \frac{[\beta(N_S + N_B)]^{N_{ON}}}{N_{ON}!} e^{-\beta(N_S+N_B)} \frac{[\beta(N'_S + N_B)]^{N_{OFF}}}{N_{OFF}!}$$

$$e^{-\beta(N'_S+N_B)} e^{-\frac{(1-\beta)^2}{2\sigma_\beta^2}}. \tag{6.11}$$

In this, β acts as a normalization factor and σ_β is the width of the Gaussian function (see, for instance, Silverwood et al. (2015); Lefranc et al. (2015); Moulin et al. (2019)). β is found by maximizing the likelihood function such that $d\mathcal{L}/d\beta \equiv 0$ and it writes as:

$$\beta(N_{ON}, N_{OFF}, N_S, N_B) = \frac{-\sigma_\beta^2(N_S + N'_S) + 1 \pm \sqrt{(\sigma_\beta^2(N_S + N'_S + 2\widehat{N_B}) - 1)^2 + 4\sigma_\beta^2(N_{ON} + N_{OFF})}}{2}. \tag{6.12}$$

The equation demonstrates the reliance of the definition of β on the measured statistics, introduced through N_{ON}, N_{OFF}, and $\widehat{N_B}$. $\widehat{N_B}$ has been derived with Eq. (6.9). The profile of β as a function of σ_β is depicted in Fig. 6.5 for a specific bin of the likelihood function and a constant DM mass of $m_{DM} = 1$ TeV. The value of $\sigma_\beta = 0.01$, representing the 1% systematic uncertainty level, is highlighted. The TS profile, computed by incorporating the Gaussian nuisance parameter for the systematic uncertainty, is illustrated in Fig. 6.6 and compared to the TS profile computed with the standard likelihood definition. The TS profile obtained with the inclusion of $\sigma_\beta = 0.03$ is also presented. Inclusion of $\sigma_\beta = 0.01$ in the TS computation leads to a 95% C.L. upper limit on the free parameter that is 20% less constraining with respect to the case where no uncertainty on the measured background is considered. For $\sigma_\beta = 0.03$, the upper limit is 50% less constraining.

6.5.2 Uncertainty About the Dark Matter Distribution

The J-factors derived from measurements are subject to both statistical and systematic uncertainties. To incorporate the statistical uncertainty in the determination of the DM distribution in the GC region, a nuisance parameter can be introduced into the likelihood. This nuisance parameter follows a log-normal distribution and can be factored into either Eq. (6.11) or Eq. (6.7). The log-normal distribution is characterized by a mean value of $\bar{J}$ and a width of σ_J (as detailed in, for example, Abdallah et al. (2020, 2021)). The log-normal distribution is expressed as:

Fig. 6.5 The profile of the β parameter concerning σ_β for $m_{\mathrm{DM}} = 1$ TeV. The gray dashed line emphasizes the value corresponding to $\sigma_\beta = 0.01$

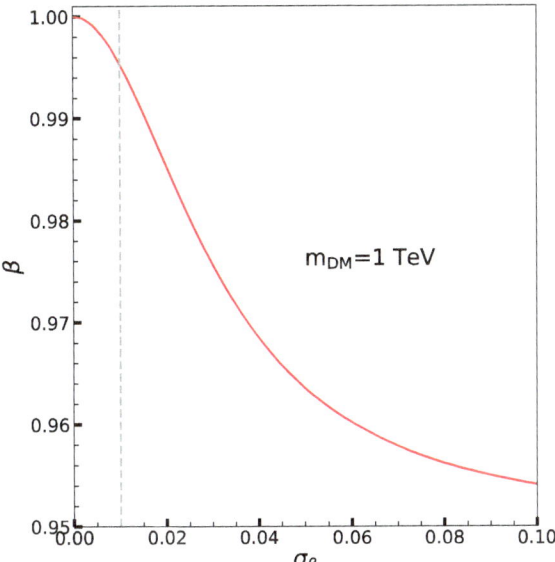

$$\mathcal{L}^J(J|\bar{J}, \sigma_J) = \frac{1}{\sqrt{2\pi} \log(10)\, \sigma_J\, J}\, \exp\left(-\frac{(\log_{10} J - \overline{\log_{10} J})^2}{2\sigma_J^2}\right). \qquad (6.13)$$

The measured J-factor represents a Gaussian realization that follows the $\mathcal{L}^J$ distribution. The optimal value of J is obtained by maximizing $\mathcal{L}^J$. The expected value is calculated as $\hat{J} = \bar{J} e^{-\sigma_J^2 \log^2(10)}$. Subsequently, $\hat{J}$ is derived and utilized to normalize the number of expected events from DM, transforming $N_S \to N_S \hat{J}/\bar{J}$. The TS profile, which includes the statistical uncertainty on the J-factor, is depicted in Fig. 6.6 and juxtaposed with the TS profile obtained from the standard likelihood definition. For this example, an arbitrary value of $\sigma_J = 0.4$ has been chosen to demonstrate how much this would degrade the final constraints. However, the determination of σ_J depends on the object harbouring the DM. The 95% C.L. upper limit on the free parameter is 2.8 times less constraining when considering the statistical uncertainty on the J-factor in the TS computation.

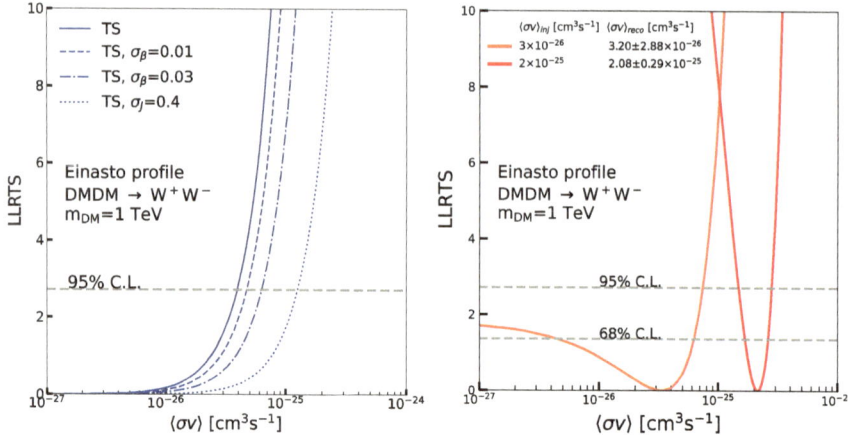

Fig. 6.6 TS profiles for a fixed DM mass, annihilation channel, and DM density profile are presented. In the left panel, the standard computation (blue solid line) is juxtaposed with profiles that incorporate the residual background uncertainty for $\sigma_\beta = 0.01$ (dashed line), $\sigma_\beta = 0.03$ (dotted-dashed line), and the statistical uncertainty on the J-factor $\sigma_J = 0.4$ (dotted line). In the right panel, TS profiles for the reconstruction of injected values of $\langle \sigma v \rangle_{inj}$ are shown. 95 and 68% C.L. are included for comparison with the depth of the TS profiles. Reconstructions at 68% C.L. (orange line) and more than 5σ (red line) are provided, along with the reconstructed value and the 68% containment bands in the legend

6.6 Performance with Fake Injected Signals

A simulated DM signal is introduced by selecting a specific mass and a designated value of $\langle \sigma v \rangle_{inj}$. This simulation involves manipulating the measured OFF distributions, which are presumed to lack a DM signal. By summing N_S and N'_S, which are obtained with assuming the chosen simulated DM signal, into the measured OFF distributions N_{OFF}, ON and OFF fake distributions are built, respectively. This procedure allows us to evaluate the framework's ability to recover the injected signal. The measured OFF distribution as previously introduced are used here. Values of $\langle \sigma v \rangle$ ranging from 3×10^{-26} to 2×10^{-25} cm^{3}s^{-1} are explored, assuming a DM mass of 1 TeV and particles annihilating in W^+W^-, with the DM distributed according to the Einasto profile. The TS procedure is executed for each injected value $\langle \sigma v \rangle_{inj}$, resulting in the computed reconstructed annihilation cross section $\langle \sigma v \rangle_{reco}$, scanning the range of annihilation cross sections cited above. The corresponding TS profiles are displayed in the right panel of Fig. 6.6. The values of $\langle \sigma v \rangle_{reco}$ and the 1 σ bands are detailed in the legend. Notably, the injection of 2×10^{-25} cm^{3}s^{-1} yields signal recovery at a 5 σ level, while for 3×10^{-26} cm^{3}s^{-1}, only the 68% containment bands are recovered.

References

Abdalla, H., et al.: Search for dark matter annihilation signals from unidentified fermi-LAT objects with H.E.S.S. Astrophys. J. **918**(1), 17 (2021). https://doi.org/10.3847/1538-4357/abff59

Abdallah, H., et al.: Search for dark matter annihilation in the Wolf-Lundmark-Melotte dwarf irregular galaxy with H.E.S.S. Phys. Rev. D **103**(10), 102002 (2021). https://doi.org/10.1103/PhysRevD.103.102002

Abdallah, H., et al.: Search for dark matter signals towards a selection of recently detected DES dwarf galaxy satellites of the milky way with H.E.S.S. Phys. Rev. D **102**(6), 062001 (2020). https://doi.org/10.1103/PhysRevD.102.062001

Ajello, M., et al.: 3FHL: the third catalog of hard fermi-LAT sources. Astrophys. J. Suppl. Ser. **232**(2), 18 (2017). https://doi.org/10.3847/1538-4365/aa8221

Cowan, G., et al.: Asymptotic formulae for likelihood-based tests of new physics. Eur. Phys. J. C **71**, 1554 (2011). https://doi.org/10.1140/epjc/s10052-011-1554-0. eprint: 1007.1727 (physics.data-an)

de Naurois, M., Rolland, L.: A high performance likelihood reconstruction of γ-rays for imaging atmospheric Cherenkov telescopes. Astropart. Phys. **32**, 231–252 (2009). https://doi.org/10.1016/j.astropartphys.2009.09.001. arXiv: 0907.2610 [astro-ph.IM]

Lefranc, V., et al.: Prospects for annihilating dark matter in the inner galactic halo by the cherenkov telescope array. Phys. Rev. **D91**(12), 122003 (2015). https://doi.org/10.1103/PhysRevD.91.122003

Li, T.-P., Ma, Y.-Q.: Analysis methods for results in gamma-ray astronomy. Astrophys. J. **272**, 317–324 (1983). https://doi.org/10.1086/161295. Sept

Lisanti, M., et al.: Mapping extragalactic dark matter annihilation with galaxy surveys: a systematic study of stacked group searches. Phys. Rev. D **97**, 063005 (6 Mar 2018). https://doi.org/10.1103/PhysRevD.97.063005. https://link.aps.org/doi/10.1103/PhysRevD.97.063005

Montanari, A., Moulin, E., Rodd, N.L.: Toward the ultimate reach of current imaging atmospheric Cherenkov telescopes and their sensitivity to TeV dark matter. Phys. Rev. D **107**(4), 043028 (2023). https://doi.org/10.1103/PhysRevD.107.043028. https://link.aps.org/doi/10.1103/PhysRevD.107.043028

Moulin, E., et al.: Science with the Cherenkov telescope array: dark matter programme. In: Science with the Cherenkov Telescope Array. World Scientific, pp. 45–81 (2019). https://doi.org/10.1142/9789813270091_0004

Silverwood, H., et al.: A realistic assessment of the CTA sensitivity to dark matter annihilation. JCAP **03**, 055 (2015). https://doi.org/10.1088/1475-7516/2015/03/055

Wilks, S.S.: The large-sample distribution of the likelihood ratio for testing composite hypotheses. Ann. Math. Stat. **9**(1), 60–62 (1938). ISSN: 00034851. http://www.jstor.org/stable/2957648

Chapter 7
Dark Matter Search in the Galactic Center

Abstract With the ingredients for searching for dark matter (DM) annihilation introduced, this chapter is dedicated to a search for a signal from self-annihilating Weakly Interacting Massive Particles in the Galactic Center region with H.E.S.S. After the presentation of the data collected with H.E.S.S. observations—the Inner Galaxy Survey observational program, some insight into the low-level details of the data taking is given. The main steps of the data analysis performed to determine whether a gamma-ray excess was present in the dataset are discussed. The spatial and spectral measured energy-count distributions for an expected dark matter signal are discussed. From the Test-Statistics analysis and its setup, the constraints obtained on $\langle \sigma v \rangle$ of the DM particles are discussed together with the impact of systematic uncertainties on the constraints.

Keywords Dark matter · Gamma rays · Galactic center · Galactic halo · Survey · Upper limits · Annihilation cross-section

7.1 Observations with H.E.S.S

H.E.S.S. observed the inner halo of the Milky Way during both Phase I and Phase II of the instrument. Both datasets consist of observational runs lasting around 28 min in the best-case scenario—when no issue interrupts the run.

Phase-I observations occurred from 2004 to 2013, directed towards pointing positions distributed around the Galactic Center (GC)—with an offset from 0.7 to 1.1°, with the specific focus to monitor the supermassive black hole Sgr A*. Selection of γ-ray events followed standard quality criteria (Aharonian et al. 2006). All observations occurred during nominal darkness conditions, with an additional requirement of an observational zenith angle less than 50° to minimize systematic uncertainties in event reconstruction. The mean zenith angle for selected observations was 19°. Data were analyzed in CT1-4 Stereo mode (de Naurois and Rolland 2009). The left panel of Fig. 7.1 illustrates the exposure map (m²s) derived from this dataset. Exposure is computed by convolving time exposure with the acceptance of the H.E.S.S. Phase I instrument, as detailed in Lefranc et al. (2015). This dataset exhibits nearly uniform

© The Author(s), under exclusive license to Springer Nature Switzerland AG 2024   139
A. Montanari and E. Moulin, *Searching for Dark Matter with Imaging Atmospheric Cherenkov Telescopes*, https://doi.org/10.1007/978-3-031-66470-0_7

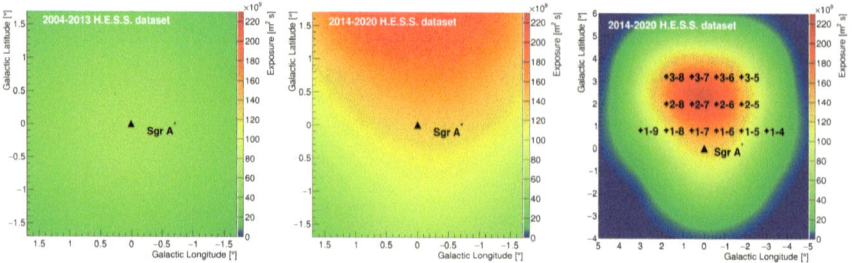

Fig. 7.1 Exposure maps of the GC region for the H.E.S.S. I and H.E.S.S. II observation phases. *Left panel*: Exposure map (in m²s) of the H.E.S.S.-I observations of the GC region (Lefranc et al. 2015). *Mid panel*: Zoomed exposure map of the H.E.S.S.-II observations of the GC region and IGS program (Abdalla et al. 2022). *Right panel*: Exposure map for the whole coverage of H.E.S.S.-II observations during the IGS program (Abdalla et al. 2022). The position of the supermassive black hole Sagittarius A* is symbolized by the black triangle

exposure within the inner $\sim |1.7|°$ of the GC region in both longitudes and latitudes. With a total of 254 h of data, this dataset was utilized for the 2016 publication on the search for dark matter (DM) annihilation signals (Lefranc et al. 2015).

The analysis presented in this chapter makes use of the observations conducted during Phase II, which are introduced in the next section.

7.1.1 The Inner Galaxy Survey

The extensive H.E.S.S. Phase II dataset comprises a total of 546 h of data, collected with observations directed towards the inner halo of the Milky Way—targeting positions near the GC and those related to the Inner Galaxy Survey (IGS). This Phase II dataset spans the period from 2014 to 2020. Same as for the Phase I dataset, the γ-ray events were selected following standard quality criteria (Aharonian et al. 2006). Observations taken in 2014 and 2015 were directed mostly towards the supermassive black hole Sagittarius A*. They were chosen for the needs of the Galactic plane survey (Abdalla et al. 2018) and dedicated source observations such as the pulsar PSR J1723-2837. The IGS was initiated in 2016 to extensively cover the GC region, specifically focusing on Galactic longitudes $|l| < 5°$ and latitudes b ranging from $-3°$ to 6°. In the right panel of Fig. 7.1, the exposure map for the entire 2014–2020 dataset is presented, obtained by convolving the time-exposure map with the acceptance of the H.E.S.S. telescopes during the IGS observations. The pointing positions of the IGS are also indicated. A minimum acceptance-corrected time exposure of 10 h is achieved up to $b \approx +6°$ with the 2014–2020 dataset, as highlighted in Fig. 7.1. A zoomed view of the exposure map obtained for this dataset in the region covered by Phase I observations is provided in the mid panel of Fig. 7.1. The increased observational time and enhanced sensitivity of the full five-telescope H.E.S.S. array contribute to approximately five times more exposure for the 2014–2020 dataset.

This Phase II dataset served as the basis for an update on the search for DM annihilation signals from the GC region (Abdalla et al. 2022) and the exploration of TeV emission from the low-latitude Fermi Bubbles (Moulin et al. 2021) with the H.E.S.S. instrument.

As shown in Abdalla et al. (2022), this dataset provides the most stringent limits on the velocity-weighted annihilation cross section of annihilating DM in the TeV mass range, obtained so far at the moment of the writing. Part of the analysis presented in Abdalla et al. (2022) is also recalled later in this chapter.

7.1.2 Low-Level Analysis of the Data Taking

7.1.2.1 Zenith and Offset Distributions

As per standard criteria adopted for Imaging Atmospheric Cherenkov Telescopes (IACTs) data selection, observations with a zenith angle lower than 45° were prioritized to minimize systematics. Nevertheless, this was not always feasible due to limited time windows, particularly in the 17–18 h right ascension band, where numerous astrophysical objects of interest are located. Consequently, the zenith angle values for the whole dataset of observations range from 3.0° to 60.0°, with a mean value of 18.0°.

In 2014 and 2015, observations were conducted on pointing positions closer to the GC, resulting in mean offset values between the nominal position of SgrA* and the pointing positions of 1.1° and 1.5°, respectively, with mean zenith angles of 19.7°.

Post-2016, observations mostly occurred at 2-X and 3-X pointing positions (see Fig. 7.1), leading to an increase in the mean offset. For the years 2016 to 2020, mean offset values of 2.2°, 3.0°, 2.7°, 2.8°, and 3.3° were recorded, respectively, along with corresponding mean zenith values of 12.0°, 13.2°, 19.7°, 18.2°, and 18.3°.

Table 7.1 summarizes the mean zenith angle values for each year in the dataset. Should the reader be interested in having more details about the per-year distributions of zenith and offset angles from the GC, please consult Sect. 6.3 in Montanari (2022).

Table 7.1 Mean zenith angles for the observational runs in each year of the 2014–2020 dataset

Years	2014	2015	2016	2017	2018	2019	2020
Mean Zenith [°]	19.7	19.7	12.0	13.2	19.7	18.2	18.3

7.1.3   Observational and Instrumental Systematic Uncertainties

The inner halo of the Milky Way presents a complex environment characterized by numerous sources emitting high-energy (HE) and very-high-energy (VHE) gamma-rays. In this region, the Night Sky Background (NSB) undergoes notable fluctuations, potentially influencing the measurement of the residual background at lower energies. The upcoming sections detail our investigation to identify potential correlations between the gamma rate in the FoV and the NSB level. The homogeneity and isotropy of the measured residual background across the FoV are explored by analyzing gamma-like rate distributions from various pointing positions in the survey. Due to the correlation of the residual background with the zenith angle of observations, an anticipated gradient in the gamma-like rate is considered. Subsequent sections outline our approach to address this effect, including the treatment of potential systematic uncertainties arising from imperfect azimuthal symmetry across the telescope's FoV. Additionally, the computation of counts is elaborated as a function of the azimuth angle to account for systematic uncertainties. The analysis extends to the assessment of energy scale uncertainty by quantifying the energy shift affecting energy reconstruction in common events across two H.E.S.S. analysis chains. While these studies are tailored to configurations necessary for DM annihilation signal search, the procedures for deriving systematic uncertainties are presented in a sufficiently general manner applicable to analyses with other datasets.

7.1.3.1   Night Sky Background and Gamma-Like Rate Correlation

Within the IGS dataset's exposure, the NSB encounters notable variations due to the presence of bright stars and diffuse emission in the inner part of the GC region and close to the Galactic plane. NSB levels range from a minimum of 25 MHz to a maximum of 400 MHz photoelectron rate per pixel in the FoV. However, it's important to note that the minimum and maximum NSB values occur in regions of the sky not covered by the region of interest for the search for DM presented in this chapter. The NSB map for this region is illustrated in Fig. 7.2. In the shower template analysis—used as the main analysis chain for this chapter, a dedicated NSB treatment is implemented (de Naurois and Rolland 2009), where the NSB contribution is modeled in every pixel of the camera. This analysis procedure eliminates the need for additional image cleaning to extract pixels illuminated by the showers. Nevertheless, a thorough examination is conducted to identify potential residual NSB and gamma-like-rate correlations.

For this assessment, squared regions around specific pointing positions of the IGS are defined, each with a 1° side and composed by squared pixels of 0.1° × 0.1°. Extracting NSB values from these pixels, distributions are constructed and mean and RMS values for the NSB rate are computed. This investigation reveals no discernible correlations between NSB and the background distribution for the various pointing

Fig. 7.2 Night Sky Background map in MHz of the inner halo of the Milky Way in Galactic coordinates. The region is zoomed to highlight the zone where maximum exposure is obtained with the dataset of H.E.S.S.-II IGS observations. Figure extracted from Montanari (2022)

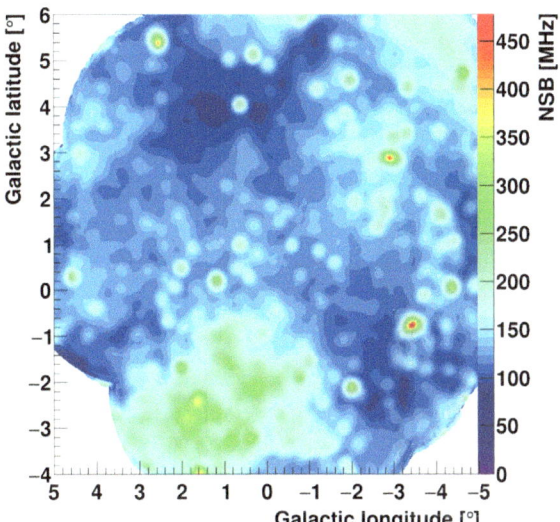

positions considered. The left panel of Fig. 7.3 illustrates an example of this procedure applied on the squared region extracted around pointing position 2–7 of the IGS dataset, where the NSB fluctuates between 110 MHz and 150 MHz. The scale of the panel deliberately focuses on the range between 100 and 160 MHz to highlight NSB fluctuations. The right panel of Fig. 7.3 depicts the gamma-like rate's evolution concerning changes in the NSB of the region.

When considering the full exposure of the H.E.S.S. II dataset, regions of the sky may be extracted where the NSB varies between approximately 100 and 300 MHz. Considering these two values, the gamma-like rate in the right panel of Fig. 7.3 exhibits variations of up to 1%. However, it's worth noting that such a significant difference in NSB rates is not observed for all pointing positions and region of interest utilized in the analysis presented in this chapter for the DM search.

7.1.3.2 Zenith Angle and Gamma-Like Rate Correlation

To assess the homogeneity of the background rate across the FoV, gamma-like number counts from the previously defined squared regions are extracted, the counts on a pixel-by-pixel basis are renormalized by time exposure, subsequently calculating mean and RMS values for the distributions. The measured RMS results are larger than what would be expected from statistics alone. Consequently, a mean systematic uncertainty value of 4% is derived from all the regions considered in this study. Figure 7.4 illustrates the measured counts in one of the squared regions extracted around pointing position 2–7 of the IGS dataset.

For analyses with the IGS dataset, where the region of interest can de defined differently, it may be useful to estimate the systematic uncertainty on a run-by-run

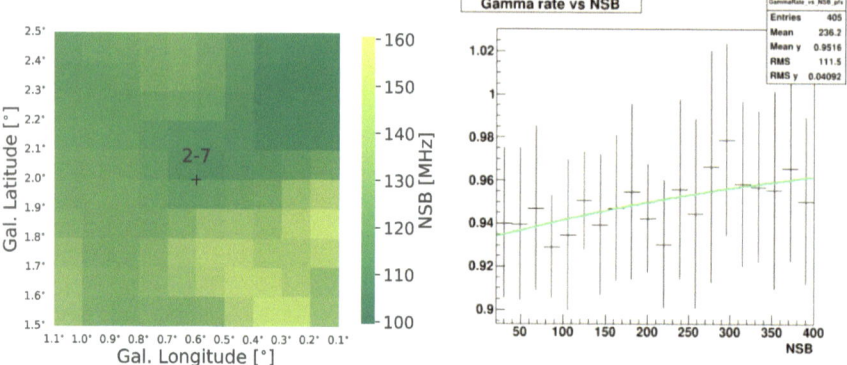

Fig. 7.3 The map of the NSB rate measured in MHz for the squared region around the IGS pointing position 2–7 is shown in the left panel. The right panel shows the gamma-like rate as a function of the NSB measured in the entire FoV of the 2014–2020 H.E.S.S. dataset. The NSB rates span from a minimum of 25 MHz up to 400 MHz, due to the high variability in the region. Figure extracted from Montanari (2022)

basis. Therefore, a second approach is adopted to investigate potential inhomogeneity of the background rate across the FoV due to the gradient in the zenith angle of observations. The correlation between the difference in zenith angle values and the gamma-like rate gradient is explored, taking the DM search analysis as an example.

In this analysis, *ON* regions, where the expected signal is searched for, and control *OFF* regions, where the residual background is measured, are defined. By applying the *Reflected Background* method, the ON region is reflected with respect to the pointing position to define the OFF region (more details about ON and OFF regions, and the Reflected Background method are presented later in Sect. 7.2). However, by construction, there are different values of zenith angles for events in the ON and OFF regions. For each degree of difference in the zenith angle across the FoV, a 1% gradient in the gamma-like rate is anticipated, as per standard measurement with the H.E.S.S. cameras (de Naurois and Rolland 2009). To test this, distributions of zenith angles per region of interest (ROI) and per pointing position are constructed. Two examples are shown in Fig. 7.5. The figure includes mean values of the distributions, along with the nominal zenith angle of the pointing position. From the test, a maximum difference between the mean values $\theta_{z,ON}$ and $\theta_{z,OFF}$ of 1° is obtained. For each run, ON and OFF distributions can be renormalized by this difference, accounting for the gradient of gamma-like rate in the FoV. However, the typical width of 1° in the obtained distributions introduces a systematic uncertainty. Considering the expected 1% gradient for each degree of difference in the zenith angle, a systematic uncertainty of 1% is considered. This value will be applied to the normalization of the event energy distributions used for subsequent analyses, and adopted with the Gaussian nuisance parameter as described in Sect. 6.5.1.

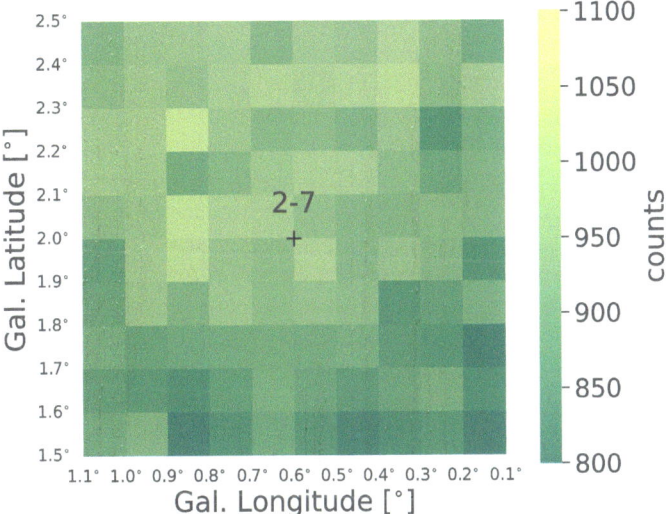

Fig. 7.4 The map shows the distribution of measured background events within a squared region. This region has a 1° side length, and its pixels are squared with dimensions of 0.1° × 0.1°. The region is centered around the IGS pointing position 2–7. Figure extracted from Montanari (2022)

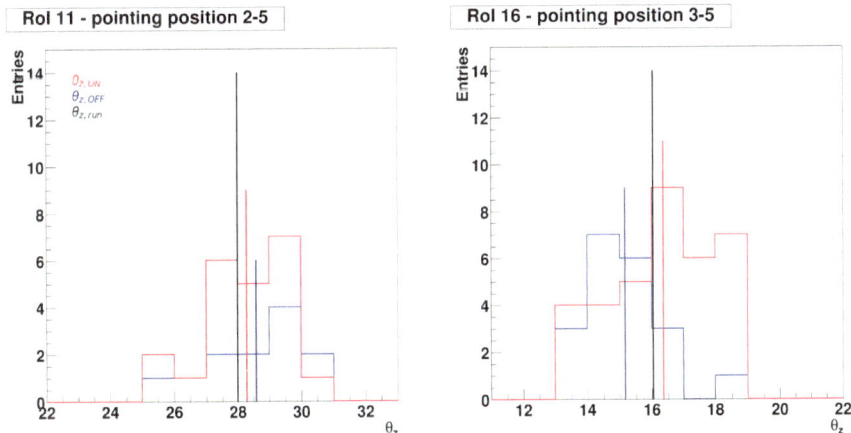

Fig. 7.5 The zenith distributions were generated for the DM search analysis with the Reflected Background method. Two regions of interest and two distinct pointing positions were considered. The red distribution represents events extracted from the ON region, while the blue distribution corresponds to the OFF region. The lines of the respective colors indicate the mean values of these distributions. Additionally, the black line represents the nominal zenith angle of the pointing position. Figure extracted from Montanari (2022)

7.1.3.3   Azimuthal Symmetry in the Field of View

To assess azimuthal symmetry in the FoV, annular regions were constructed around selected pointing positions. A specific example is presented for a ring with inner radius $r_{in} = 0.7°$ and outer radius $r_{out} = 0.8°$. This choice of dimensions reflects the typical offset between the source and the pointing position during H.E.S.S. observations. The left panel of Fig. 7.6 illustrates an instance of such a ring in Galactic coordinates, centered on the IGS pointing position 2–6.

To investigate azimuthal symmetry and identify potential preferred angles in the camera FoV, the rings were divided into 36 angular bins. Gamma-like rates were then estimated for each bin, and a distribution over the bins was constructed to extract mean and RMS values. No systematic uncertainty was identified, except for the anticipated 1% per degree of zenith angle gradient in the FoV. The right panel of Fig. 7.6 displays the counts for each of the 36 angular bins. A sinusoidal function, $f(\alpha) = p_0 + p_1 \sin(k\alpha + p_2)$, was fitted to the data, testing for the first harmonic $(k = 1)$. The fit results indicate that p_1 is compatible with zero, suggesting the absence of a first harmonic. Consequently, the distribution of counts over the angular bins aligns well with a constant, indicating no preferred angle in the camera FoV.

7.1.3.4   Uncertainty on the Energy Scale

In the H.E.S.S. collaboration, two main analysis chains are used for the energy reconstruction of gamma-ray candidates. A discrepancy is found when the two reconstruc-

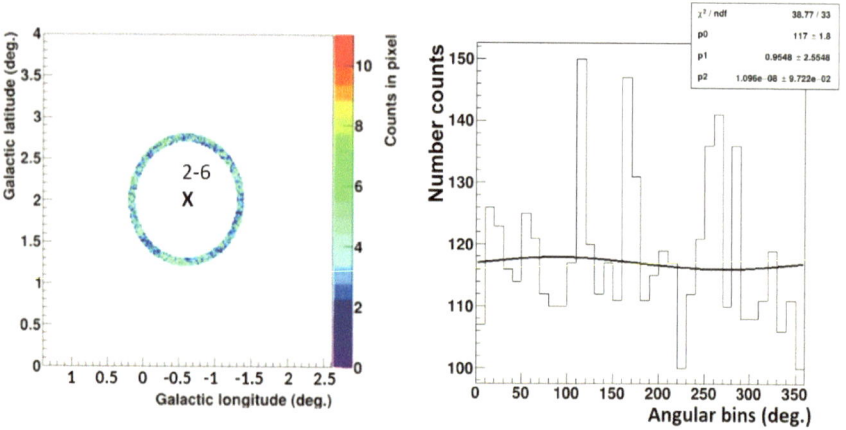

Fig. 7.6 Test for azimuthal symmetry on one pointing position of the IGS dataset. *Left panel*: annular region built around pointing position 2–6 for $r_{in} = 0.7°$ and $r_{out} = 0.8°$, in galactic coordinates. The color scale shows the count in each pixel of $0.02° \times 0.02°$. *Right panel*: fit of the number counts extracted from 36 angular bins with the function defined as $f(\alpha) = p_0 + p_1 \sin(k\alpha + p_2)$. The fit for $k = 1$ is considered, i.e. the first harmonic is tested. Figure extracted from Montanari (2022)

tions are compared. A more detailed study of this comparison is presented in Sect. 6.3 in Montanari (2022). For reference, the reconstruction in the two chains revealed a systematic uncertainty of 10% on the energy scale of the energy distributions, for the IGS dataset. For the purpose of the DM search analysis, where measured count distributions are compared to expected count distributions from a simulated DM signal, it is important to note that this uncertainty equally affects the energy scale of both distributions. Consequently, it would introduce an overall shift on the energy scale of the constraints that could be computed with the dataset. As a result, no correction for this discrepancy was applied in the analysis presented in this chapter.

7.2 Data Analysis

This section explains the standard steps for setting up the search for DM signals. As a standard approach with IACT data, one first looks at excess and significance sky maps to identify potential sources of background when searching for a DM signal. Section 7.2.1 shows the maps obtained with standard methods adopted in the H.E.S.S. collaboration. Then, the ROI for the DM search has to be defined. Exclusion regions can also be placed on known VHE sources to avoid background contamination in the ROI. This is explained in Sect. 7.2.2. Once this step is completed, the residual background can be measured, as explained in Sect. 7.2.3. Finally, one should look for a potential excess in the ROI compatible with a DM signal. The search for the excess is explained in Sect. 7.2.4.

7.2.1 Excess and Significance Sky Maps

For the H.E.S.S.-II dataset, standard gamma-ray excess and significance maps are generated using the *Ring Background* (Berge et al. 2007) method in both CT1-4 Stereo and CT1-5 Stereo modes, incorporating the full five-telescope array. Notably, no standard exclusion region is applied to known VHE sources.

The resulting maps, displayed in Fig. 7.7, focus on the region between $-4°$ and $6°$ in Galactic latitudes and $|5°|$ in Galactic longitudes. In the left panel, the photon excess relative to the background is depicted, with the map artificially capped at 1000 counts. The middle panel showcases the significance map in standard deviations, capped at 15 σ. Notably, the map reveals several significant hotspots, featuring a significance of 4 σ above the background. Recognizable VHE sources such as HESS J1745-200 (Sgr A*), HESS J1747-281 (G09+01), and HESS J1745-30 are evident, along with the TeV emission from Sgr B2. Diffuse emission around the GC region is also visible. Known sources and significant hotspots will be covered with exclusion regions in the later analysis. No indications of potential source detections are identified in sky regions outside the designated masks.

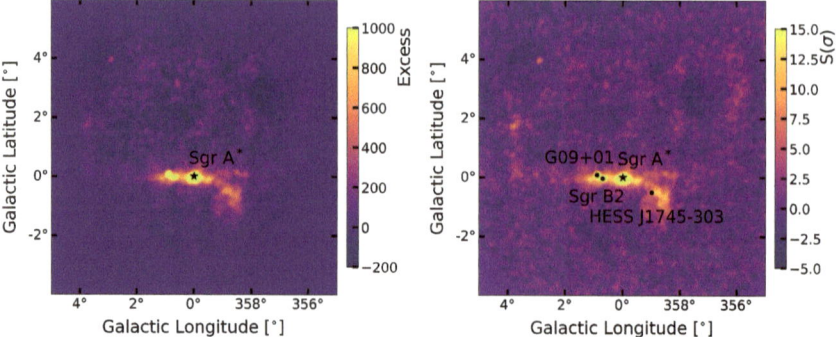

Fig. 7.7 The set of panels comprises the gamma-ray excess map (left panel) and significance map (left panel) for the H.E.S.S. II dataset. These maps are generated using the Ring Background technique in both CT1-5 Stereo and CT1-4 Stereo modes, and no exclusion region is implemented in these maps. Figure extracted from Montanari (2022)

7.2.2   Definition of the Region of Interest and Exclusion Regions

The region designated for the search for DM signals serves as the ROI, referred to as the ON region. The definition of this ROI is primarily influenced by the adopted expected DM density profile and the distribution of pointing positions across the sky. For the DM analysis, the Einasto density profile (Springel et al. 2008) is used to describe the distribution of Weakly Interacting Massive Particles (WIMPs) DM. This profile predicts a DM density that peaks in proximity to the GC. The adopted DM distribution, characterized by the J-factor (refer to Sect. 3.3), is displayed in the left panel of Fig. 7.8. The color scale in the figure represents the values of the J-factor computed for the Einasto profile within pixel dimensions of $0.02° \times 0.02°$. Considering the anticipated spatial distribution of DM, the ROI is defined as a disk centered on the nominal GC position with a radius of $3°$. To capitalize on the distinctive spatial morphology of DM signals compared to the residual background, the disk is further subdivided into 25 ROIs represented as rings with inner radii ranging from $0.5°$ to $2.9°$. Each ring has a fixed width of $0.1°$. The arrangement of these rings is depicted in the right panel of Fig. 7.8, which shows the time exposure map derived from the exposure map in the right panel of Fig. 7.1.

A set of conservative masks is adopted in the analysis to avoid VHE gamma-ray contamination in the signal and background regions. All the sources in the H.E.S.S. Galactic Plane Survey (Abdalla et al. 2018) and other possible VHE sources, marked as hotspots through the low-level H.E.S.S. data selection, are masked. The Galactic plane is masked within $0.3°$ in Galactic latitude. For pointlike sources, a circular mask of $0.25°$ radius is used. For the extended source HESS J1745-303, a circular mask of $0.9°$ radius is used. The set of mask is shown in both panels of Fig. 7.8. The first ring is created for inner radius of $0.5°$ because the masks used in the analysis exclude the inner part of the region. Without masks, one could consider even rings closer to the GC.

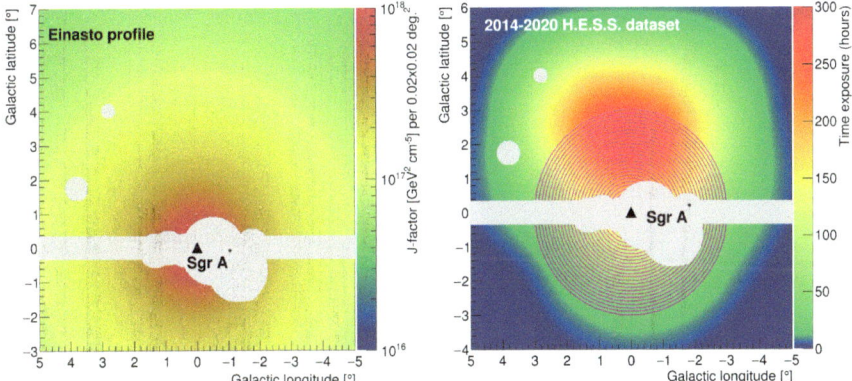

Fig. 7.8 The map in the left panel illustrates the values of the J-factor corresponding to the Einasto profile in Galactic coordinates. The integration of the J-factor is computed within pixels of dimensions $0.02° \times 0.02°$. The left panel illustrates the time exposure map of the 2014–2020 data. The region of interest for the search of DM signals is shown as the purple rings. Both panels show the gray-shaded area as the set of masks implemented in the analysis to prevent contamination from astrophysical background sources in the ROI. The position of the supermassive black hole Sagittarius A* is denoted by the black triangle in both panels. The left panel is extracted from Abdalla et al. (2022)

7.2.3 Measurement of the Residual Background

The determination of the residual background in the search for DM employs the Reflected Background method (Berge et al. 2007), as detailed. This method ensures that the background measurement is conducted simultaneously with the signal measurement in the same field of view on a run-by-run basis. The OFF region, used for background measurement, is symmetrically positioned with respect to the ON region, maintaining similar observational and instrumental conditions, as outlined in Lefranc et al. (2015); Abdalla et al. (2022); Moulin et al. (2019). Consequently, the measurements from the OFF regions are acquired under the similar enough conditions as the ON regions, allowing for a precise determination of the residual background. Although an accurate determination of the observational condition differences between the ON and OFF regions could be investigated, it is outside the scope of this book. Exclusion regions are removed consistently for both ON and OFF measurements, ensuring the same solid angle size. This procedure is carried out on a run-by-run basis, producing a reliable determination of the background.

An illustrative example of the construction of the OFF region for background measurement is presented in Fig. 7.9 for ROIs 7, 13, and 25, along with the respective pointing positions (indicated by black crosses) 2–5 ($l = -1.8°$, $b = 2.0°$) and 3–7 ($l = 0.8°$, $b = 3.2°$). The exclusion regions are consistently subtracted from both the ON and OFF measurements, ensuring identical solid angle size and acceptance. The color scale depicts the same J-factor distribution as shown in Fig. 7.8. Notably, a significant expected DM excess signal in the ON region, relative to the OFF region, is evident, as indicated by the ratios between the J-factor values in the ON and OFF regions presented in the figure. For ROI 13, the ratios between the J-factor in the ON and OFF regions are 5 and 4, respectively, concerning the pointing positions 3–7 and 2–5.

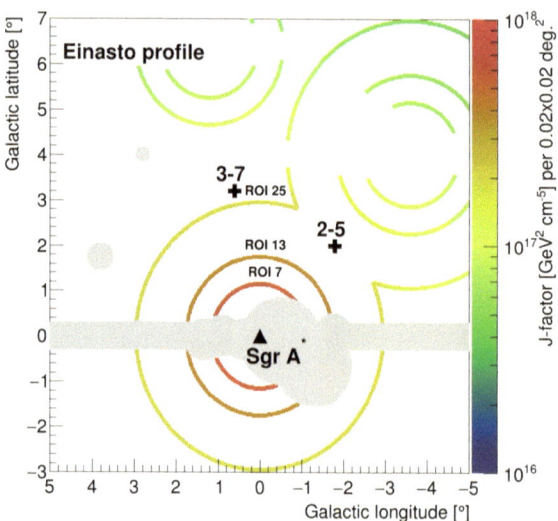

Fig. 7.9 The background measurement method is presented on a J-factor map in Galactic coordinates for two distinct pointing positions of the IGS (3–5 and 3–7), denoted by black crosses in the figure. J-factor values are indicated via the color bar for ROIs 7, 13, and 25, along with the corresponding values obtained in the corresponding OFF regions. The DM distribution is assumed to follow an Einasto profile. The supermassive black hole Sagittarius A* position is marked by the black triangle for reference. The grey shaded region corresponds to the masked region. Figure extracted from Abdalla et al. (2022)

7.2.4 Search for a Gamma-Ray Excess

Following the reconstruction of spectral and spatial information as gamma-ray-like events, each event is categorized based on the ring of the ROI in which it was observed. Event distributions are then constructed for each ring of the defined ROI, showing the number of events as an energy function.

Photon statistics in the ON and OFF regions, obtained from the energy count distributions for each ring of the ROI, along with the excess significance, are provided in Table 7.2. Photon statistics and excess significance are reported for energy bins above the safe energy thresholds of E_{th} = 200 GeV. The excess significance is determined using Eq. (6.10). Should the reader be interested in a more detailed inspection of the obtained energy count distributions, please consult Sect. 8.2 in Montanari (2022).

The energy-differential spectra, derived from the event energy distributions in the ON and OFF regions for selected ROI rings and the combination of all rings, are illustrated in Fig. 7.10. Despite a thorough analysis, no significant excess is observed in any of the ROIs. Please note that the energy-differential spectra shown in Fig. 7.10 are obtained with the H.E.S.S. instrument response functions and the specific geometry adopted for the H.E.S.S. DM search analysis. The latter depends on the ROI definition, the application of the masks, and the run-by-run information, which modify the ON and OFF region construction for each observation in the IGS dataset. Therefore,

Table 7.2 Photon statistics and excess significance for each of the 25 ROIs. The first row indicates the ROI number. The second and third rows present the measured photon statistics in the ON and OFF regions, respectively, above the energy threshold. The fourth row reports the excess significance calculated using the ON and OFF statistics with Eq. (6.10). Figure extracted from Abdalla et al. (2022)

ith ROI	1	2	3	4	5	6	7	8	9	10	11	12
N_{ON}	326	1830	3029	4736	6793	9144	12036	15201	16830	19530	23549	25585
N_{OFF}	298	1674	3087	4665	6699	9164	11899	15177	17242	19721	23270	25568
$S(\sigma)$	1.1	2.6	−0.7	0.7	0.8	−0.2	0.9	0.1	−2.2	−0.9	1.3	0.1
13	14	15	16	17	18	19	20	21	22	23	24	25
27571	29875	32328	35094	37292	39957	42540	42460	42282	42317	42653	43188	42879
27673	29945	32518	34774	37502	40159	42775	42939	42415	42509	42896	43011	43373
−0.4	−0.3	−0.8	1.2	−0.8	−0.7	−0.8	−1.6	−0.5	−0.7	−0.8	0.6	−1.7

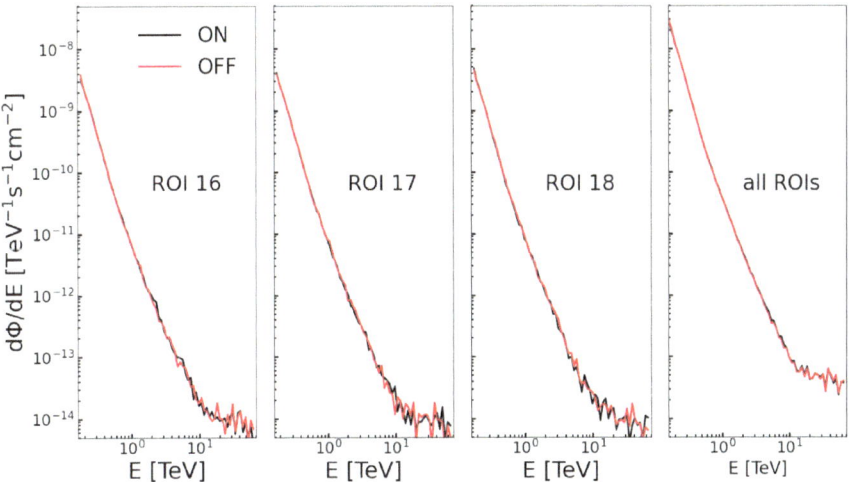

Fig. 7.10 Energy-differential spectra obtained for ON (black lines) and OFF (red lines) regions for ROI 16, 17 and 18, respectively, in the first three panels. The energy-differential spectra for the ON and OFF regions of the combination of all the ROI rings are shown in the last panel. Figure extracted from Abdalla et al. (2022)

the spectra cannot be considered as a direct estimate of the H.E.S.S. flux sensitivity level for the GC region. Unfortunately, since the instrument response functions and the software used for the H.E.S.S. analysis are proprietary at the moment of writing, we cannot divulge this information. Therefore, without this information, one can not expect to strictly reproduce the H.E.S.S. analysis.

The search for significant DM signals involves comparing the ON and OFF event energy distributions for each ring, and the resulting excess significance is calculated using Eq. (6.10). The excess significance, expressed in terms of σ, is reported in the third row of Table 7.2 for each ROI. No significant excess compatible with the spatial and spectral features of the searched DM signal was found in the ROI. The reader

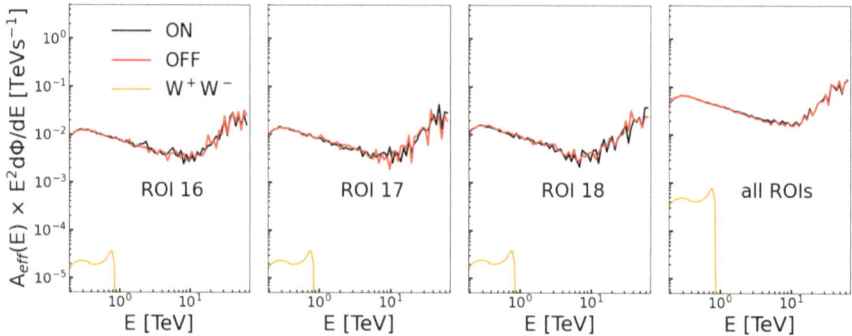

Fig. 7.11 Expected energy-differential spectra from self-annihilating DM particles with a mass $m_{DM} = 0.98$ TeV and $\langle \sigma v \rangle = 3.8 \times 10^{-26}$ cm^{3}s^{-1} in the W^+W^- annihilation channel. The spectra are presented in E^2 and are convolved with the H.E.S.S. response (orange line) for individual ROIs as well as for the combination of all ROIs. $A_{\rm eff}(E)$ represents the energy-dependent acceptance of the instrument. Additionally, the ON (black line) and OFF (red line) energy-differential spectra are plotted. The first three panels depict the spectra for individual ROIs, while the last panel shows the spectra for the combination of all ROIs. Figure extracted from Abdalla et al. (2022)

interested in more details on the analysis to extract the significance values in each ring should refer to Abdalla et al. (2022), and Sect. 8.2 of Montanari (2022).

7.3 Expected Annihilation Signals

7.3.1 Dark Matter Profile

The DM distribution in the GC region is typically parameterized following the Einasto and NFW profiles, which expressions were given in Eqs. (3.8) and 3.9, respectively. The DM density at the solar position is assumed as $\rho_\odot = 0.39$ GeV cm^{-3} (Catena and Ullio 2009). The parameters used in this analysis for these two profiles, and an alternative parameterization for the Einasto profile were already provided in Table 3.1.

An example of the J-factor map for the Einasto profile was already shown in Fig. 7.8. For each ring in the ROI, the total J-factor values are computed. The J-factor values for the Einasto and NFW profiles are shown in Table 7.3. The second and third columns provide the inner and outer radii for each ring. The solid angle is given in the fourth column. The last three columns provide the J-factor values for the three profiles tested in this analysis. The J-factor obtained for the Einasto parameterization assumes at least a factor 2 larger values than from the other two parameterizations, therefore, more constraining limits are expected with the former.

Table 7.3 J-factor values in the ROI considered in this analysis, shown in units of GeV^{2}cm^{-5}. The ring number, the inner and the outer radii, and the solid angle size for each ring are given in the first four columns. The J-factor values in the rings, computed without applying the masks on the excluded regions, are given for the Einasto, an NFW (Abdallah et al. 2016) and an alternative Einasto (Cirelli et al. 2011) profiles in the fifth, sixth and seventh columns, respectively. Figure extracted from Abdalla et al. (2022)

ith ROI	Inner radius [deg.]	Outer radius [deg.]	Solid angle $\Delta\Omega$ [10^{-4} sr]	J-factor $J(\Delta\Omega)$ [10^{20} GeV^{2}cm^{-5}]		
				Einasto	NFW	Einasto (Cirelli et al. 2011)
1	0.5	0.6	1.05	9.5	4.9	3.0
2	0.6	0.7	1.24	9.8	4.9	3.2
3	0.7	0.8	1.44	10.1	4.9	3.3
4	0.8	0.9	1.63	10.2	4.8	3.4
5	0.9	1.0	1.82	10.3	4.8	3.5
6	1.0	1.1	2.01	10.4	4.8	3.5
7	1.1	1.2	2.20	10.5	4.7	3.6
8	1.2	1.3	2.39	10.5	4.7	3.6
9	1.3	1.4	2.58	10.5	4.7	3.6
10	1.4	1.5	2.77	10.5	4.6	3.7
11	1.5	1.6	2.97	10.4	4.6	3.7
12	1.6	1.7	3.16	10.4	4.6	3.7
13	1.7	1.8	3.35	10.3	4.5	3.7
14	1.8	1.9	3.54	10.3	4.5	3.7
15	1.9	2.0	3.73	10.2	4.5	3.7
16	2.0	2.1	3.92	10.2	4.5	3.7
17	2.1	2.2	4.11	10.1	4.4	3.7
18	2.2	2.3	4.31	10.0	4.4	3.7
19	2.3	2.4	4.50	9.9	4.4	3.7
20	2.4	2.5	4.69	9.9	4.3	3.6
21	2.5	2.6	4.88	9.8	4.3	3.6
22	2.6	2.7	5.07	9.7	4.3	3.6
23	2.7	2.8	5.26	9.6	4.3	3.6
24	2.8	2.9	5.45	9.5	4.3	3.6
25	2.9	3.0	5.64	9.5	4.2	3.6

7.3.2 Energy-Differential Fluxes in the Region of Interest

The energy-differential annihilation spectrum in the W^+W^- channel, considering the convolution with the acceptance and the energy resolution for H.E.S.S., for a

Fig. 7.12 Expected DM events for a dark matter mass of $m_{DM} = 3$ TeV, annihilation channel W^+W^-, and annihilation cross section $\langle \sigma v \rangle = 5 \times 10^{-24}$ cm^{3}s^{-1} as the orange distribution. For comparison, the ON and OFF event energy distributions for ROI 12 are displayed with 1σ error bars for each energy bin. Figure extracted from Montanari (2022)

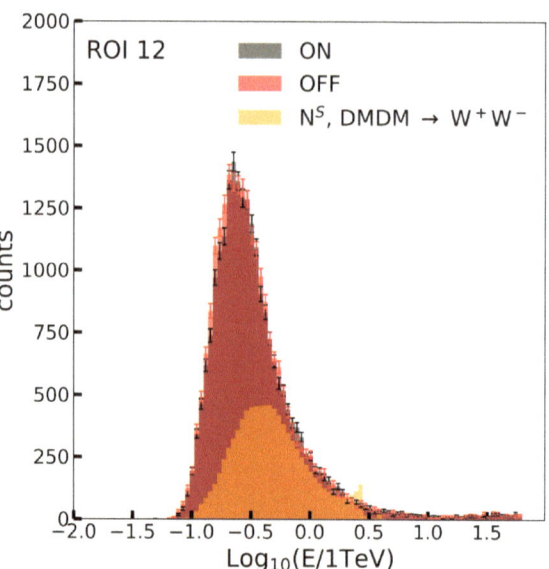

DM particle mass of $m_{DM} = 0.98$ TeV[1] and annihilation cross section of $\langle \sigma v \rangle = 3.8 \times 10^{-26}$ cm^{3}s^{-1}, is depicted in Fig. 7.11. The spectrum is presented for individual ROIs as well as for the combination of all ROIs. Overlaid on the plot are the energy-differential spectra for the corresponding ON and OFF regions, convolved with the same H.E.S.S. acceptance and energy resolution. The averaged H.E.S.S. acceptance, extracted for the observations in the IGS dataset, was shown in the right panel of Fig. 5.9. The ON and OFF spectra exhibit an increase after the break at around 10 TeV, attributed to events reconstructed with only the four small telescopes, which are more sensitive at higher energies. To compute the expected number of photons from annihilating DM, considering a fixed DM particle mass and annihilation channel, the energy differential flux per spectral and spatial bin defined in Eq. (3.7) is used. This computation incorporates instrument response functions, such as the effective area, energy resolution, and observational live time. The expected number of photons from annihilating DM, denoted as N_S, is obtained by summing $N_{S,k}$ over all runs k. The energy resolution of the H.E.S.S. telescopes is considered via the convolution of the spectrum with a Gaussian of σ/E of 10% above 200 GeV. The energy resolution is represented by $R(E_\gamma, E'\gamma)$, relating the detected energy $E'\gamma$ to the true energy E_γ of the events. The expected number of photons includes the J-factor $J(\Delta\Omega)$ for an ROI of solid angle $\Delta\Omega$, the energy-dependent acceptance of the instrument $A_{\text{eff},k}(E_\gamma)$ for run k, and the observational live time $T_{\text{obs},k}$ for each run. The expression for $N_{S,k}$ for self-annihilating Majorana DM particles of mass m_{DM} in the channels f is given by:

---

[1] This specific value of the DM particle mass has been chosen for the sole reason that it was selected among the H.E.S.S. energy bins available for the analysis.

$$N_{S,k}(\langle\sigma v\rangle) = \frac{\langle\sigma v\rangle J(\Delta\Omega)}{8\pi m_{DM}^2} T_{obs,k} \int_{E_{th}}^{m_{DM}} \int_0^\infty \sum_f BR_f \frac{dN_\gamma^f}{dE_\gamma}(E_\gamma) R(E_\gamma, E'_\gamma) A_{eff,k}(E_\gamma) dE_\gamma dE'_\gamma, \quad (7.1)$$

Figure 7.12 shows the event energy distribution for DM annihilating into the W^+W^- channel, for a DM particle mass of $m_{DM} = 3$ TeV and annihilation cross section of $\langle\sigma v\rangle = 5 \times 10^{-26}$ cm^{3}s^{-1}, and for ROI 12 in the analysis. This is obtained with Eq. (7.1), including all the runs in the dataset. Overlaid on the plot are the event energy distributions for the corresponding ON and OFF regions.

7.4 Constraining a Dark Matter Signal with the Test Statistics

Once it has been determined that no excess compatible with the spatial and spectral features expected for the searched DM signal, one can use the Test Statistics (TS) framework defined in Sect. 6.3 to obtain limits on the parameters of the adopted DM model.

In the context of the analysis presented, as no significant excess compatible with DM signals is observed in the ROI, upper limits on the annihilation cross section $\langle\sigma v\rangle$ are derived. These limits are determined using the statistical framework defined in Sect. 6.3 and following Cowan et al. (2011).

In the statistical analysis, a 2-dimensional log-likelihood ratio test statistic is employed, taking into account the expected spectral and spatial features of dark matter signals. This analysis is conducted in 67 logarithmically-spaced energy bins and 25 spatial bins corresponding to the rings of the ROI. The safe energy threshold at 200 GeV is used, excluding the computation of limits for masses below this threshold. The likelihood function used is the same as the one defined in Sect. 6.2, where $N_{S,i,j}$ and $N'_{S,i,j}$ represent the total number of dark matter events in the (i, j) spatial and spectral bins for the ON and OFF regions, respectively. These values are computed using Eq. (7.1), taking into account the energy-dependent acceptance and energy resolution. The gamma-ray yield term dN_γ^f/dE_γ, corresponding to the gamma-ray yield in channel f, is calculated using the Monte Carlo event collision generator PYTHIAv8.135, including final state radiative corrections (Cirelli et al. 2011). To incorporate systematic uncertainties into the likelihood function, a Gaussian nuisance parameter is introduced, consisting of $\beta_{i,j}$ as a normalization factor and $\sigma_{\beta,i,j}$ as the width of the Gaussian function (see, for instance, Silverwood et al. (2015); Lefranc et al. (2015); Moulin et al. (2019) and Sect. 7.1.3 for the estimate of the uncertainty). The value of $\beta_{i,j}$ is determined by maximizing the likelihood function, ensuring that $d\mathcal{L}_{i,j}/d\beta_{i,j} \equiv 0$. The $\sigma_{\beta,ij}$ value is fixed to 1%. The results including this uncertainty are discussed in Sect. 7.5.

The analysis assumes a positive signal $\langle\sigma v\rangle > 0$. In the high statistics regime, the TS follows a χ^2 distribution with one degree of freedom. The derived limits exclude values of $\langle\sigma v\rangle$ with TS higher than 2.71 at 95% C.L.

The results of this analysis are shown in Sect. 7.4.1. Finally, Sects. 7.4.2 and 7.4.3 compare the results obtained in this analysis with limits obtained with other experiments and for different assumptions of the DM density distribution in the GC region. At the moment of the writing, the limits on $\langle \sigma v \rangle$ of the DM particles, for the tested annihilation channels, are the most stringent ones for DM masses in the TeV range.

7.4.1 Expected and Observed Limits on the Annihilation Cross Section

For the upper limits on $\langle \sigma v \rangle$ for self-annihilation of WIMPs with DM masses ranging from 200 GeV up to 70 TeV, different annihilation channels are considered, including quark ($b\bar{b}$, $t\bar{t}$), gauge bosons ($W^+ W^-$, ZZ), lepton ($e^+ e^-$, $\mu^+ \mu^-$, $\tau^+ \tau^-$) and Higgs (HH). The observed and expected upper limits at 95% C.L. for the $W^+ W^-$ and $\tau^+ \tau^-$ channels, respectively, for the mentioned Einasto profile are shown in Fig. 7.13. The 68 and 95% statistical containment bands are also plotted. The observed limits are computed using the available statistics in the ON and OFF measured energy count distributions.

In the $W^+ W^-$ channel, the observed upper limit on the annihilation cross section for a DM particle with a mass of 1.5 TeV is 3.7×10^{-26} cm^{3}s^{-1}. For the $\tau^+ \tau^-$ annihilation channel, the obtained upper limit is 1.2×10^{-26} cm^{3}s^{-1} for a DM mass of 0.7 TeV.

At the moment of the writing, an improvement factor of 1.6 is achieved for a DM particle with a mass of 1.5 TeV with respect to the latest constraints in Abdallah et al.

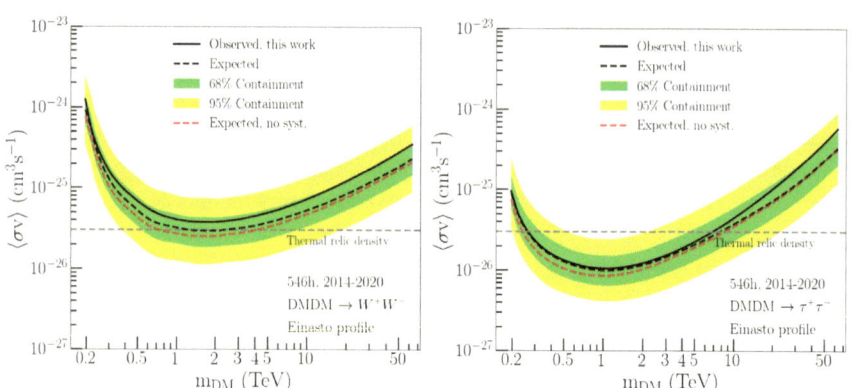

Fig. 7.13 Upper limits on $\langle \sigma v \rangle$ as a function of the DM mass m_{DM}, derived from the 2014–2020 H.E.S.S. observations for the $W^+ W^-$ (left panel) and $\tau^+ \tau^-$ (right panel) channels. The upper limits include the systematic uncertainty. Observed upper limits are represented by the black solid line. Mean expected upper limits (black dashed line) along with the 68% (green band) and 95% (yellow band) C.L. statistical containment bands are also displayed. The mean expected upper limits computed without including systematic uncertainty are shown as the red dashed line. The natural scale expected for thermal-relic WIMPs is indicated by the horizontal gray long-dashed line. The Figure was extracted from Abdalla et al. (2022)

(2016). The increased statistics from the dataset, collected over a longer observational live time, and the deployment of the CT1-5 array of H.E.S.S. contributed to the enhanced sensitivity of the present analysis.

The limits for the W^+W^- and $\tau^+\tau^-$ annihilation channels are shown in Fig. 7.13. The interested reader should refer to Abdalla et al. (2022) for the limits for the other annihilation channels ($b\bar{b}$, $t\bar{t}$, ZZ, HH, e^+e^-, and $\mu^+\mu^-$). The values of $\langle\sigma v\rangle$ expected for DM particles with thermal-relic cross section (Bertone et al. 2005) are intersected by the limits in the $\tau^+\tau^-$ and e^+e^- annihilation channels.

7.4.2 Comparison with Other Ongoing Experiments

The limits obtained with this analysis are the most constraining ones at the moment of the writing in the TeV mass range. In Fig. 7.14, a comparison is shown between the limits obtained in this analysis and previous limits from H.E.S.S. observations

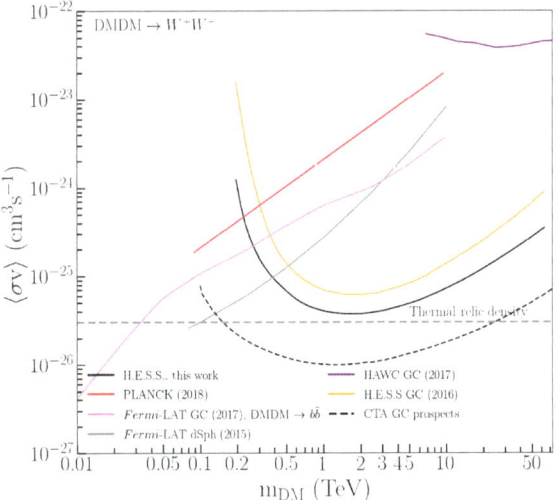

Fig. 7.14 Comparison between upper limits on $\langle\sigma v\rangle$ for the W^+W^- channel, obtained with DM indirect detection techniques. The H.E.S.S. limits from the IGS program are shown as the solid black line (Abdalla et al. 2022). The H.E.S.S. limits from observations of the GC from Abdallah et al. (2016) are shown as an orange line. Limits from GC observations with HAWC are presented as a purple line (Abeysekara et al. 2018). Additionally, limits from the observations of 15 dwarf galaxy satellites of the Milky Way by the Fermi satellite (Ackermann et al. 2015) and from the GC region (Ackermann et al. 2017) are shown as the gray and violet lines, respectively. The Fermi limits are shown for the $b\bar{b}$ channel. Furthermore, limits from the cosmic microwave background with PLANCK are included as the red line (Aghanim et al. 2018). Additionally, prospects of observations of the GC region with CTA are shown as the black dashed line (Acharyya et al. 2021). The Einasto profile is used for all the GC limits. The Figure was extracted from Abdalla et al. (2022)

and other experiments in the GeV-TeV mass range, specifically for the W^+W^- annihilation channel.

The previous H.E.S.S. limits from observations of the GC region, obtained with the H.E.S.S.-I dataset of 254 h of observations, are included (Abdallah et al. 2016). Limits from observations of the GC region with HAWC (Abeysekara et al. 2018) and *Fermi*-LAT (Ackermann et al. 2017) are also shown. Additionally, the *Fermi*-LAT limits from observing 15 dwarf galaxy satellites of the Milky Way are also displayed (Ackermann et al. 2015). The *Fermi*-LAT limits are presented for the $b\bar{b}$ annihilation channel. Furthermore, the limits obtained from the cosmic microwave background measured by PLANCK (Aghanim et al. 2018) are included in the comparison.

The presented limits from this analysis are 1.6 times more constraining than the H.E.S.S. constraints from Abdallah et al. (2016) for a DM particle with a mass of 1.5 TeV. Additionally, they surpass the limits obtained with *Fermi*-LAT for particles with masses above approximately 300 GeV.

7.4.3 Testing Different Dark Matter Density Profiles

In Fig. 7.15, the results for the different DM density profiles adopted for this analysis are compared. The limits computed with the J-factor values obtained with different profiles are shown. The profiles include NFW and Einasto (with 2 parameterizations).

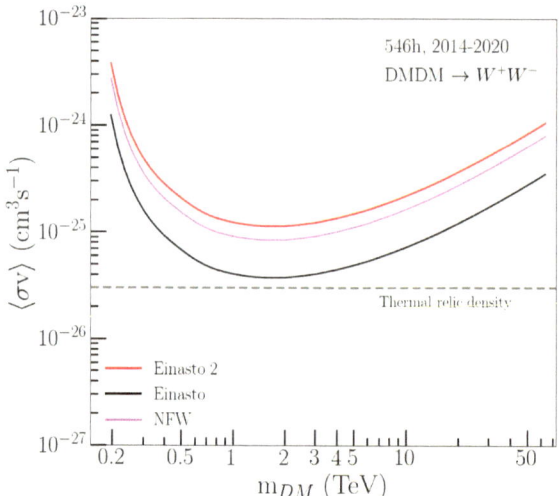

Fig. 7.15 Limits on the velocity-weighted annihilation cross section $\langle \sigma v \rangle$ are presented, obtained with different parameterizations of the dark matter density distribution. The 95% confidence level upper limits are shown for the Einasto profile (black line), another parameterization of the Einasto profile (Cirelli et al. 2011) referred to as "Einasto 2" (red line), and the NFW profile (pink line). These limits are computed for the W^+W^- channel and include the systematic uncertainty. The figure illustrates the impact of different density profiles on the derived constraints, emphasizing the sensitivity of the results to the assumed dark matter distribution. The Figure was extracted from Abdalla et al. (2022)

As described in Sect. 3.3, computing the limits with the NFW or the Einasto 2 parameterizations results in about a factor of 2.5 weaker constraints compared to the ones obtained with the Einasto profile. If a DM density distribution is assumed to be kiloparsec-sized cored, such as the Burkert profile, the limits would be weakened by about two orders of magnitude. Conversely, assuming a Moore-like profile would produce more constraining limits by about a factor of two.

This comparison highlights the impact of the assumed dark matter density profile on the derived constraints and emphasizes the importance of accurate modeling in DM searches.

7.5 Impact of the Systematic Uncertainties

In the systematic study for this analysis, the focus is on the uncertainty associated with the normalization of the energy count distributions. The derivation involves considering the expected gradient in the gamma-like rate based on the difference in the zenith angle of the observations. This was explained in Sect. 7.1.3.

Specifically, for this DM analysis, the difference in the mean values of the distributions of the ON and OFF event zenith angles is examined for each chosen ring of the ROI and each pointing position of the IGS. The observed gradient of the gamma-ray-like rate in the FoV is then renormalized based on the difference in the mean values of the zenith angle distributions for the ON and OFF datasets on a run-by-run basis.

The typical width of $1°$ of the zenith angle distribution is taken into account by introducing a systematic uncertainty of 1% on the normalization of the measured energy count distributions. This uncertainty is incorporated as a Gaussian nuisance parameter, which includes $\beta_{i,j}$ as a normalization factor and $\sigma_{\beta,ij}$ as the width of the Gaussian function. Here, $\sigma_{\beta,ij}$ is fixed to 1%.

Including this systematic uncertainty has the effect of deteriorating the mean expected limits, causing them to worsen from 8 to 18% depending on the DM particle mass. Figure 7.13 shows the observed limits with the inclusion of this systematic uncertainty. The expected limits are shown for two cases: with and without the uncertainty. It's important to note that, for this particular analysis, no other sources of systematic uncertainties are considered.

References

Abdalla, H., et al.: Search for dark matter annihilation signals in the H.E.S.S. inner galaxy survey. Phys. Rev. Lett. **129**(11), 111101 (2022). https://doi.org/10.1103/PhysRevLett.129.111101

Abdalla, H., et al.: The H.E.S.S. galactic plane survey. Astron. Astrophys. **612**. A1 (2018). https://doi.org/10.1051/0004-6361/201732098

Abdallah, H., et al.: Search for dark matter annihilations towards the inner Galactic halo from 10 years of observations with H.E.S.S. Phys. Rev. Lett. **117**(11), 111301 (2016), p. 111301. https://doi.org/10.1103/PhysRevLett.117.111301

Abeysekara, A.U., et al.: A search for dark matter in the Galactic halo with HAWC. JCAP **2018**(02), 049–049 (2018). https://doi.org/10.1088/1475-7516/2018/02/049

Acharyya, A., et al.: Sensitivity of the Cherenkov telescope array to a dark matter signal from the galactic centre. J. Cosmol. Astropart. Phys. **2021**(01), 057 (2021). https://doi.org/10.1088/1475-7516/2021/01/057

Ackermann, M., et al.: Searching for dark matter annihilation from milky way dwarf spheroidal galaxies with six years of fermi large area telescope data. Phys. Rev. Lett. **115**(23), 231301 (2015). https://doi.org/10.1103/PhysRevLett.115.231301

Ackermann, M., et al.: The fermi galactic center GeV excess and implications for dark matter. Astrophys. J. **840**(1), 43 (2017). https://doi.org/10.3847/1538-4357/aa6cab

Aghanim, N., et al.: Planck 2018 results. VI. cosmological parameters. Astron. Astrophys. **641** (2020). [Erratum: Astron. Astrophys. **652**, C4 (2021)], A6. https://doi.org/10.1051/0004-6361/201833910

Aharonian, F., et al.: Observations of the crab nebula with H.E.S.S. Astron. Astrophys. **457**, 899–915 (2006). https://doi.org/10.1051/0004-6361:20065351

Berge, D., Funk, S., Hinton, J.: Background modelling in very-high-energy γ-ray astronomy **466**(3), 1219–1229 (2007). https://doi.org/10.1051/0004-6361:20066674. https://doi.org/10.1051/0004-6361:20066674

Bertone, G., Hooper, D., Silk, J.: Particle dark matter: evidence, candidates and constraints. Phys. Rept. **405**, 279–390 (2005). https://doi.org/10.1016/j.physrep.2004.08.031. eprint: `hep-ph/0404175`

Catena, R., Ullio, P.: A novel determination of the local dark matter density. JCAP **08**, 004 (2010). https://doi.org/10.1088/1475-7516/2010/08/004

Cirelli, M., et al.: PPPC 4 DM ID: a poor particle physicist cookbook for dark matter indirect detection. JCAP **1103**, 051 (2011). https://doi.org/10.1088/1475-7516/2012/10/E01, https://doi.org/10.1088/1475-7516/2011/03/051

Cowan, G., et al.: Asymptotic formulae for likelihood-based tests of new physics. Eur. Phys. J. C **71**, 1554 (2011). https://doi.org/10.1140/epjc/s10052-011-1554-0. eprint: `1007.1727` (physics.data-an)

de Naurois, M., Rolland, L.: A high performance likelihood reconstruction of γ-rays for imaging atmospheric Cherenkov telescopes. Astropart. Phys. **32**, 231–252 (2009). https://doi.org/10.1016/j.astropartphys.2009.09.001. arXiv: 0907.2610 [astro-ph.IM]

Lefranc, V., et al.: Prospects for annihilating dark matter in the inner galactic halo by the Cherenkov telescope array. Phys. Rev. **D91**(12), 122003 (2015). https://doi.org/10.1103/PhysRevD.91.122003

Montanari, A.: Study of the galactic center and search for dark matter with the H.E.S.S. inner galaxy survey. 2022UPASP098. Ph.D. thesis (2022). http://www.theses.fr/2022UPASP098/document

Moulin, E., et al.: Science with the Cherenkov telescope array: dark matter programme. In: Science with the Cherenkov Telescope Array. World Scientific, pp. 45–81 (2019). https://doi.org/10.1142/9789813270091_0004

Moulin, E., et al.: Search for TeV emission from the fermi bubbles at low galactic latitudes with H.E.S.S. inner galaxy survey observations. PoS ICRC2021, 791 (2021). arXiv: 2108.10028 [astro-ph.HE]

Silverwood, H., et al.: A realistic assessment of the CTA sensitivity to dark matter annihilation. JCAP **03**, 055 (2015). https://doi.org/10.1088/1475-7516/2015/03/055

Springel, V., White, S.D.M., Frenk, C.S., et al.: A blueprint for detecting supersymmetric dark matter in the Galactic halo. Nature **456**, 73–76 (2008)

Chapter 8
Sensitivity Reach to TeV Dark Matter

Abstract Even though TeV-scale dark matter (DM) candidates face challenges from decades of null searches, the scenario remains compelling given that simple realizations such as Wino and Higgsino DM remain undetected. The ultimate sensitivity, at the moment of the writing, of current Imaging Atmospheric Cherenkov Telescopes is explored in the context of a broad range of TeV-scale DM candidates—including specific ones such as the Wino, Higgsino, and Quintuplet. Realistic mock H.E.S.S.-like observations of the inner Milky Way halo are performed and several uncertainties impacting the limits are explored—from theoretical expectations on the DM distribution and spectra, to the instrumental and astrophysical background uncertainties. H.E.S.S.-like instruments can obtain results competitive with neutrino telescopes when exploring annihilation into neutrino line final states. The sensitivity to the Wino and Quintuplet can probe thermal masses, while the thermal Higgsino is still standing a factor of a few out of reach for the adopted DM distributions.

Keywords Dark matter · Gamma rays · Galactic halo · TeV DM models · Astrophysical backgrounds · Sensitivity prospects

8.1 Probing TeV Dark Matter Models in the Galactic Center

Chapter 3 already introduced some of the theoretical models one can assume for the computation of the gamma-ray flux from dark matter (DM) annihilation. However, these assumptions introduce a degree of uncertainty, given that no wide consensus is reached—at the moment of the writing—on the spectral shape of a putative DM annihilation signal or on the spatial morphology of the DM distribution in several astrophysical environments observed for DM searches.

This section aims at introducing some alternatives to the models previously assumed. Gamma-ray spectra from DM annihilation obtained with the `HDMSpectra` public software (Bauer et al. 2021) will be presented. We will make use of Milky Way DM mass profiles obtained with mass-modeling measurements of the Milky Way

© The Author(s), under exclusive license to Springer Nature Switzerland AG 2024   161
A. Montanari and E. Moulin, *Searching for Dark Matter with Imaging Atmospheric Cherenkov Telescopes*, https://doi.org/10.1007/978-3-031-66470-0_8

rotation curve. These alternative models will be employed to obtain the sensitivity reach of the current generation of Imaging Atmospheric Cherenkov Telescopes (IACTs) to DM annihilation signals, utilizing simulated datasets on the basis of the measured Inner Galaxy Survey (IGS) data presented in Chap. 7.

Considering the ongoing effort in the community to improve the description of the expected DM spectra and spatial morphology, the aim here is not to be exhaustive but to present a general-enough landscape of what the current reach of IACTs to DM annihilation signals is at the moment of the writing.

8.1.1 PPPC4DMID and HDMSpectra Gamma-Ray Yields

The public code `PPPC4DMID` (Cirelli et al. 2011) is widely used for the computation of the gamma-ray yield. A more recent one is also considered, `HDMSpectra` (Bauer et al. 2021). The readers interested in an in-depth understanding of the theoretical modelling used to compute these yields is referred to the provided references. The comparison for spectra of photons from self-annihilating Weakly Interacting Massive Particles (WIMPs) into the W^+W^- channel, computed with the `PPPC4DMID` and `HMDSpectra` softwares, was shown in the right panel of Fig. 3.5. Additionally, the gamma-ray yield for self-annihilation into the three neutrino line channels, $\nu_\mu\bar{\nu}_\mu$, $\nu_e\bar{\nu}_e$ and $\nu_\tau\bar{\nu}_\tau$, is presented for further comparison with limits obtained with ANTARES later in this chapter. It's important to note that final state neutrinos produced from DM annihilation may emit W and Z gauge bosons, leading to the generation of continuous gamma-ray spectra (Queiroz et al. 2016). The spectra of DM particles self-annihilating in these three neutrino channels are illustrated in Fig. 8.1, using the `HDMSpectra` yield, for DM masses $m_{\mathrm{DM}} = 20, 40, 60, 80$ and 100 TeV.

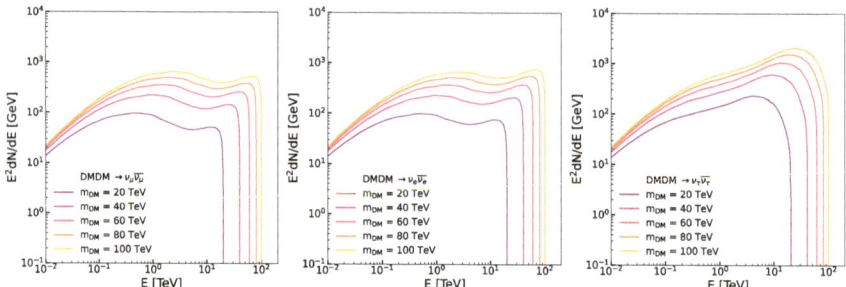

Fig. 8.1 Spectra of photons expected for WIMPs self-annihilation into the $\nu_\mu\bar{\nu}_\mu$, $\nu_e\bar{\nu}_e$ and $\nu_\tau\bar{\nu}_\tau$ channels. Spectra for DM masses of $m_{\mathrm{DM}} = 20, 40, 60, 80$ and 100 TeV obtained from the `HDMSpectra` (Bauer et al. 2021) software package are shown

8.1.2 The Wino, Higgsino, and Quintuplet Models

In what was discussed so far, DM annihilation signal searches were always performed in model-independent frameworks. Despite the evolving landscape of theoretical and experimental results, exploring DM signals with annihilation cross sections around the value leading to the correct relic abundance remains a valuable and model-independent avenue for searches across a wide range of final states. That said, there remains a strong motivation to explore more specific realizations of WIMPs. One avenue is introducing minimal field content to the Standard Model to explain DM. This can be achieved by incorporating TeV-scale states charged under the electroweak interaction, including an SU(2) doublet with unit hypercharge, as well as **3** and **5** representations of SU(2) (Cirelli et al. 2006, 2007, 2008; Cirelli and Strumia 2009; Cirelli et al. 2015; Mahbubani and Senatore 2006; Kearney et al. 2017). These states are known as the Higgsino, Wino, and Quintuplet, respectively. They can account for the total observed DM abundance through thermal production in the early universe for masses of 1 ± 0.1, 2.9 ± 0.1, and 13.6 ± 0.8 TeV, respectively (Cirelli et al. 2007; Hisano et al. 2007; Hryczuk et al. 2011; Beneke et al. 2016; Mitridate et al. 2017; Bottaro et al. 2022a).

Detection prospects for these minimal DM candidates are broadly discussed, see, for instance, Bottaro et al. (2022a), Bottaro et al. (2022b). Higgsino and Wino also serve as thermal DM candidates that consistently realize supersymmetry in accordance with observations from the LHC (Arkani-Hamed et al. 2012; Fox et al. 2014; Hall et al. 2013). Although the precise paths for discovering DM in these scenarios remain unclear (Co et al. 2022a,b), the possibility of detecting signals from Higgsino dark matter with the Cherenkov Telescope Array (CTA) (see Sect. 5.6 for an introduction on CTA) provides strong motivation for assessing the existing sensitivity of ACTs (Rinchiuso et al. 2021). In this chapter, the sensitivity of IACTs to Higgsino, Wino, and Quintuplet models will also be assessed. These models are fully defined, and their particle physics contributions to the gamma-ray yield are completely specified. The thermal masses of these models are fixed to specific values, eliminating all free parameters (except for the choice of two mass splittings for the Higgsino). Evaluating the full mass range that can be covered by IACT is relevant in case the early Universe deviated from the standard thermal relic cosmology.

Each of these WIMP scenarios could annihilate into a two-photon final state, providing a gamma-ray line at the DM mass as a feature target. However, determining the cross-section and gamma-ray yield for these models requires considering Sommerfeld enhancement, resummation effects of order m_{DM}/m_W, and additional channels beyond direct annihilation into two photons. The Wino model accounts for these effects, and the analysis uses a next-to-leading logarithmic (NLL) computation (Baumgart et al. 2019). The Quintuplet model has been recently extended with the same formalism (Baumgart et al. 2023).

For the Higgsino model, a similar comprehensive computation is not yet available, so the approach from Rinchiuso et al. (2021) is applied. This involves including the leading-order (LO) computation of the line and continuum, along with Sommerfeld

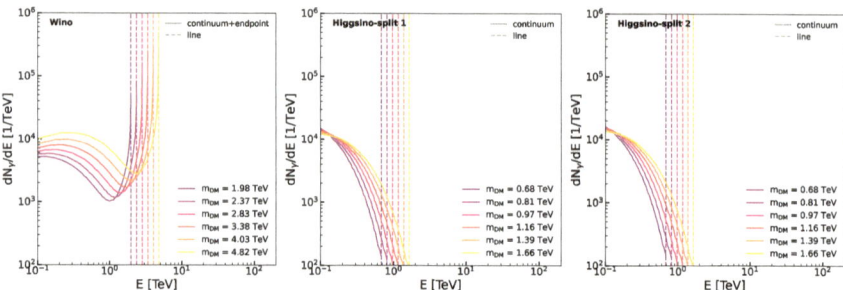

Fig. 8.2 Theoretical gamma-ray yields expected for self-annihilating Winos (left panel) Higgsinos in split 1 (central panel) and 2 (right panel). The spectra show the continuum and endpoint, and continuum-only contributions for Wino and Higgsino, respectively. The spectra for line contribution only, which are pure delta functions, are shown too (dashed lines). Figure extracted from Montanari et al. (2023)

enhancement. Additionally, the Higgsino model requires specifying an additional parameter, the splitting between the charged and neutral states in the spectrum, denoted as δm_+ and δm_N, respectively. Two benchmarks are chosen for this purpose: for splitting one, $\delta m_+ = 350$ MeV and $\delta m_N = 200$ keV, saturating the limits set by direct detection; for splitting two, these values are inverted to $\delta m_+ = 480$ MeV and $\delta m_N = 2$ GeV. Alternative parameterizations for the Higgsino are also available from Beneke et al. (2020). The theoretical spectra of photons associated with the continuum and endpoint contributions for Winos at different m_{DM} are presented in the left panel of Fig. 8.2. The spectra include both continuum and endpoint contributions, and lines at m_{DM} are shown as pure delta functions. Spectra of photons for the continuum contribution for several m_{DM} for Higgsinos in splits 1 and 2 are shown in the central and right panels of Fig. 8.2, along with the line contributions, respectively.

8.1.3 Uncertainties on the Dark Matter Distribution in the Milky Way

As introduced in Sect. 3.3.1, in order to infer the DM distribution in the inner part of the Milky Way, two commonly used approaches are DM(-only) cosmological simulations and mass-modeling. DM-only simulations predict cuspy DM distributions, commonly parameterized by the NFW or Einasto profiles. These profiles were used to derive the constraints on DM in Chap. 7. When baryonic physics and feedback processes are considered in simulations, the DM distribution can dynamically evolve, potentially leading to the creation of kpc-sized cores depending on the modeling of baryonic physics.

In this chapter, the Einasto (Springel et al. 2008) profile with the same parameterization applied in Chap. 7 for limits computation will be considered. At the same time, the NFW parameterization as derived from Cautun et al. (2020) will be extracted. And limits when applying the cNFW profile will be computed, which considers a core of radius $r_c = 1$ kpc. The J-factor for these three profiles have been showed in Fig. 3.4. Moreover, IACT sensitivity for two-body neutrino annihilation channels will be compared to the results obtained from ANTARES in Albert et al. (2019). To make the comparison more direct, the NFW parameterization adopted in that work will be labeled as aNFW. aNFW considers a local density $\rho_\odot = 0.47$ GeV/cm^3 and scale radius $r_s = 16.1$ kpc.

8.2 Conventional Astrophysical Background

8.2.1 Models for the Cosmic-Ray Fluxes in the Galactic Center

In the introduction to the IACT technique, it was discussed that the dominant background contribution for observations with Cherenkov telescope arises from combined fluxes of hadrons, electrons, and positrons incident on the atmosphere. These fluxes sum up to be significantly larger than the observed rate for photons from even the brightest steady very-high-energy (VHE) sources. Even though the showers generated by cosmic rays (CRs) and gamma rays can be distinguished, a residual contribution in the measured gamma-like flux is irreducible because the finite discrimination between CRs and gamma rays. The reader can consult Chap. 5 for more details.

As opposed to what was done in Chap. 7, where observations were used in the OFF regions to get a measure of the residual background, Bernlöhr et al. (2013) is followed here to compute the expected fluxes of cosmic-ray hadrons—dominated by protons and helium nuclei, as well as electrons and positrons. The events from misidentified CR-generated showers that are simulated are then defined as the residual background. For distances within $\lesssim 1$ kpc of the solar neighborhood, a spatial feature in the arrival direction of VHE CRs can be left by CR electron and positron sources; however, no anisotropy has been detected so far (Abdollahi et al. 2017). Therefore, spatial isotropy for the residual background is assumed.

8.2.2 Models for the Galactic Diffuse Emission at TeV Energies

When energetic CRs interact with interstellar material and ambient photon fields, they generate a diffuse flux of gamma rays known as the Gallactic diffuse emission

(GDE). In particular, the GDE is a combination of photons arising from neutral pions decay—produced from CR proton collisions with interstellar gas, Bremsstrahlung from these same protons, and finally the inverse Compton scattering (ICS) of CR electrons.

The GDE contributes to the majority of photons detected by $Fermi$-LAT in its energy range ($\sim$MeV–TeV) (Ackermann et al. 2012). The present uncertainty still affecting available GDE models represents a fundamental systematic error for several DM analyses conducted with $Fermi$-LAT data. However, this is not yet the case for H.E.S.S. (Abdalla et al. 2022; Montanari et al. 2023a; Montanari et al. 2023b). The GDE has yet to be conclusively detected at TeV energies at the moment of the writing.

Nevertheless, in order to investigate the ultimate IACT sensitivity to DM signals, the GDE can emerge as an important background contribution at TeV, therefore, a GDE model is included in the analysis presented in this chapter. When CTA data from the inner Galactic Center (GC) survey will be available, it will be mandatory to build realistic GDE models to maximize the utility of the increased flux sensitivity expected from this array; this point is discussed in Silverwood et al. (2015), Lefranc et al. (2015), Moulin (2017), Rinchiuso et al. (2021). As the purpose here is to consider the additional possible contribution from the GDE to an analysis with simulated data, rather than confront real data, a simplified model of the GDE will be sufficient. In particular, for this simplified approach, the "GDE scenario 2" developed in Rinchiuso et al. (2021) is adopted. The spatial distribution of the π^0 and Bremsstrahlung emission is derived from the assumption that these are tracers of the interstellar dust, as shown in Schlegel et al. (1998). A simple parametric model for the ICS from Su et al. (2010) is adopted. Then, the energy distribution of the model is fitted to the measured $Fermi$-LAT GDE data extracted from Ackermann et al. (2017). The reader is referred to Rinchiuso et al. (2021) for the more complete details.

Despite the simplified approach adopted in this chapter, one can use more advanced tools to model the expected TeV-scale diffuse gamma-ray emission. For instance, the CR propagation framework GALPROP (see Porter et al. (2022) for the complete discussion on version 57). GALPROP assumes different realistic CR source density and interstellar radiation field distributions through several sets of models. Despite the differences among these models, they agree with extensive collections of locally measured CR data and therefore can be considered representative of uncertainties related to the transport of CRs in the Galaxy.

For a given GALPROP simulation, the necessary inputs are the density distribution of CR sources, the interstellar gas distribution, interstellar radiation fields, and magnetic fields dictating the energy losses and secondary particle production. A detailed discussion of all the inputs is outside the scope of this section, therefore the interested reader is referred to Montanari et al. (2023b) for an example of all the necessary ingredients to perform such simulations and include GALPROP simulations in a DM search analysis. Two of the maps used in the analysis presented in Montanari et al. (2023b) are shown in Fig. 8.3. The left panel shows the flux maps for the hadronic component (π_0 decay) at an energy of 10 TeV, obtained with the CR source density $SA100$ and the interstellar radiation field model $R12$ (Montanari

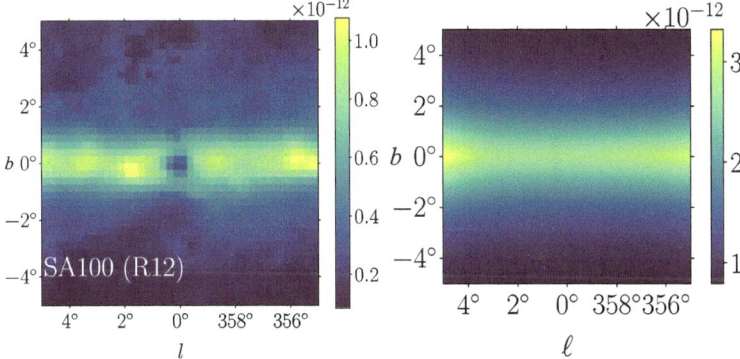

Fig. 8.3 *Left panel*: Hadronic flux map (as $d^2\Phi/dE/d\Omega$) in Galactic coordinates (l, b) for the GALPROP framework version 57 expressed for units of $\mathrm{TeV}^{-1}\ \mathrm{cm}^{-2}\ \mathrm{s}^{-1}\ \mathrm{sr}^{-1}$. The map is shown at en energy of 10 TeV, for CR source density model $SA100$ and the ISRF model $R12$. *Right panel*: Inverse Compton flux map for the same models and shown with the same units adopted for the Hadronic flux map. Figure extracted from Montanari et al. (2023b)

et al. 2023b). As expected, the spatial morphology is mostly driven by the target material density. The right panel shows the flux map for a simulated ICS component, at an energy of 10 TeV and for the same CR source density and ISRF models. As the energy of the CR $e^{\pm}$ increases, the Klein-Nishina effects become increasingly important, and the spatial morphology of the IC map reflects the CR sources spatial distribution. The Figure was extracted from Montanari et al. (2023b).

8.2.3 Gamma Rays from Millisecond Pulsars in the Galactic Bulge

An excess of gamma rays emerging from the GC has been detected by *Fermi-LAT* Goodenough and Hooper (2009), Hooper and Goodenough (2011), Ajello et al. (2016) and is commonly referred to as the Galactic Center Excess (GCE). While its nature remains under debate (see Leane et al. (2022) for a recent discussion), it may be caused by a population of millisecond pulsars (MSP) in the inner galaxy. Electrons accelerated in the wind regions of magnetospheres of pulsars could escape the pulsar environment and undergo ICS on ambient radiation fields. This would produce VHE gamma rays. Therefore, this additional ICS gamma-ray background to our analyses would need to be included.

Constraining the spectral index of the injection spectrum of $e^{\pm}$ from pulsars is challenging. The magnetic reconnection in the equatorial current sheet outside the pulsar light cylinder produces the most energetic $e^{\pm}$ (Cerutti et al. 2016). The polar cap region close to the pulsar magnetosphere is thought to be responsible for the generation of the pulsed emission. Additional uncertainty is expected on the maximum energy of the emitted $e^{\pm}$—it could reach PeV energies (Guépin et al. 2020). Bearing these caveats in mind, the emission spectrum presented in Macias et al. (2021) is considered. This can be roughly represented as a power-law spectrum $E^{-2.5}$, with an exponential cut-off at 1 TeV. The Boxy Bulge distribution described in Macias et al. (2021) is considered for the spatial morphology. Although considerable uncertainties on both the gamma-ray spectrum and morphology of the MSP contribution are still present at the moment of the writing, this contribution will be shown to have only a minor impact on IACT DM analyses. Therefore, these uncertainties will not be propagated through to the next steps in the analysis.

8.2.4 Very-High-Energy Emission from the Fermi Bubbles

As introduced in Sect. 4.4.3, for Galactic latitudes higher than 10°, the Fermi Bubbles (FBs) show a power-law energy spectrum, scaling as E^{-2}. The spectrum softens considerably above 100 GeV. For Galactic latitudes closer to the plane, brighter and harder emission from the FBs has been detected in Fermi-LAT data (Ackermann et al. 2017; Storm et al. 2017; Herold and Malyshev 2019). In particular, the emission exhibits a power-law spectrum that persists until $\sim$1 TeV. The limited photon statistics available from *Fermi*-LAT above 100 GeV obstruct any strong claims on the spectrum at higher energies and the spectrum may remain hard in the TeV energy range.

In order to model the FBs emission as a background for our DM search analysis, the best-fit spectrum above 100 GeV is extracted from Moulin et al. (2021). This was obtained by exploiting the H.E.S.S. IGS observations and showed that the FBs spectrum stays hard until the TeV energy range Moulin et al. (2021). Although the final H.E.S.S. results on the FBs are being finalized at the moment of the writing to be made public by the H.E.S.S. collaboration, this modeling is accurate enough for our purposes. For the spatial distribution of the emission, energy independence is assumed, and a spatial template derived from Herold and Malyshev (2019) is used.

8.2.5 The Galactic Center Pevatron

Finally, the H.E.S.S. Pevatron in the GC Abramowski et al. (2016) is included as the last conventional emission that one needs to account for as background when searching for a DM signal. The emission of the GC Pevatron from H.E.S.S. measurements in Abramowski et al. (2016) is presently restricted within the inner $\sim$75 pc of the GC, which corresponds to an angular scale of $\sim$0.5° (Abramowski et al. 2016); nevertheless it will be considered.

8.2.6 Expected Backgrounds and Dark Matter Signals

As introduced in the previous sections, the expected overall background in this search for DM signal is modeled by known sources of residual background and conventional emissions. CR protons and nuclei entering the atmosphere produce hadronic showers, some of which might be misidentified as gamma-ray showers due to finite rejection power. To account for this, the expected number of events produced by a flux of CR protons and helium nuclei, as well as electrons and positrons, is defined following the approach described in Bernlöhr et al. (2013). A constant rejection factor of 10 is considered for protons and helium nuclei, as reported in literature references (Rinchiuso et al. 2021; de Naurois and Rolland 2009; Bernlohr 2008).

The number of signal events $N_{S,i,j}$ in the ith region of interest (ROI) and jth energy bin for a given DM annihilation channel and density profile can be computed using the Eq. (3.7) and following the approach described in Sect. 7.4. The definition of the ROI as rings strictly follows what was already discussed in Sect. 7.2.2. The number of background events $N_{B,i,j}$ is defined. To compute this, the following substitution is made in Eq. (3.7): $d\Phi^{DM}/dE \times A^{\gamma}_{eff}$ by $d\Phi^{CR}/dE \times A^{CR}_{eff} + d\Phi^{Conv}/dE \times A^{\gamma}_{eff}$. Here, $d\Phi^{CR}/dE$ is the flux of cosmic rays, and $d\Phi^{Conv}/dE$ is the flux of conventional gamma-ray background.

The flux of cosmic rays is dictated by what is explained in Sect. 8.2.1, and fluxes of conventional gamma-ray backgrounds are produced by the sources that have been described in the Sects. 8.2.2, 8.2.3, 8.2.4, and 8.2.5. The energy-dependent acceptance for the hadronic (proton, helium) cosmic ray flux is given by $A^{CR}_{eff} = \epsilon^{CR} A^{\gamma}_{eff}$, where ϵ^{CR} is the cosmic ray efficiency. The flux of photons from the residual background is modeled from protons, helium, and electrons as power laws. For the first two spectra, the fluxes are defined as $d\Phi(E)/dE = N \times (E/1TeV)^k$, and a more complex function is adopted for the electrons: $d\Phi(E)/dE = N \times (E/1TeV)^k + L/(E\omega\sqrt{2\pi}) \exp(-(\ln(E/E_p))^2/2\omega^2)$. The parameters of the spectra are reported in Table 8.1. Here the fraction of hadronic cosmic rays ϵ^{CR} that remain identified as gamma-rays, is assumed to be 10% over the full energy range considered. This factor accounts for the finite rejection power of hadronic showers in the γ-ray observations. With ϵ^{CR} set at 10%, a photon efficiency higher than 95% can be achieved, as given in de Naurois and Rolland (2009). For this analysis, the gamma-ray acceptance for observations with the full five-telescopes H.E.S.S. array has been extracted from Holler et al. (2016). This information ensures a realistic description of the IGS observations, even though the acceptance is not strictly the same. A refined descrip-

Table 8.1 Parameterizations for the fluxes of CR spectra of protons, electrons and Helium used for background modelling, as extracted from Rinchiuso et al. (2021)

Particle	N [1/TeV m^2 s sr]	k	L	E_p [TeV]	ω
p	0.096	−2.70			
He	0.0719	−2.64			
e	6.85×10^{-5}	−3.21	3.19×10^{-3}	0.107	0.776

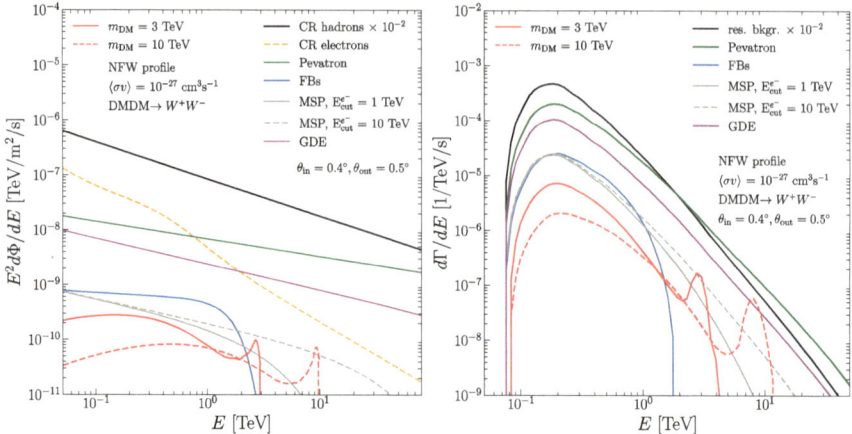

Fig. 8.4 *Left panel*: Expected spectra from WIMPs self-annihilating in the W^+W^- channel, for DM masses of $m_{DM} = 3$ and 10 TeV and with a velocity-weighted annihilation cross section $\langle \sigma v \rangle = 1 \times 10^{-27}$ cm^{3}s^{-1}. Cosmic ray fluxes for hadrons (proton + helium) (solid black line) and electrons (orange line) are plotted too. Four conventional astrophysical emissions are shown: the diffuse fluxes from the H.E.S.S. Pevatron (Abramowski et al. 2016) (green line), the base of the Fermi Bubbles Moulin et al. (2021), the expectation from the MSP-bulge population for two different values of the cut-off energy for the electron IC emission (Macias et al. 2021), and the GDE from the "GDE scenario 2" extracted from Rinchiuso et al. (2021). All the energy-differential gamma-ray fluxes are given for ROI 2. *Right panel*: Energy-differential count rates as a function of energy for signal and background in ROI 2. The Figure was extracted from Montanari et al. (2023a)

tion of the instrument response function would require dedicated simulations of both the instrument and the observations, which is beyond the scope of this study. For the purposes of this analysis, a homogeneous time-exposure of 500 h is assumed to represent what has been achieved with the IGS dataset.

The differential count rates are defined as in Eq. (3.7), for each considered emission in ROI i, by:

$$\frac{d\Gamma_{S,B,i,j}}{dE} = \frac{dN_{S,B,i,j}}{T_{obs,i}\,dE}. \tag{8.1}$$

The left panel of Fig. 8.4 shows the expected fluxes in our second ROI—so for CR, and the above mentioned conventional astrophysical emissions, *i.e.*, the GDE, the PeVatron, FBs and MSPs. The expected signal generated by DM particles of mass $m_{DM} = 3$ TeV or 10 TeV and annihilating into W^+W^- with $\langle \sigma v \rangle = 10^{-27}$ cm^{3}s^{-1}, for two different DM profiles (NFW and cNFW) is also added. The right panel of Fig. 8.4 shows the rates for the same components (Montanari et al. 2023a).

8.3 Sensitivity to TeV Annihilating Dark Matter

8.3.1 Setup for the Statistical Analysis

The region of interest for the DM search defined in Chap. 7 was limited to the innermost $\sim 3°$ of the Milky Way. For the analysis presented in this chapter, a search strategy is built upon the same example. Although, observations out to larger latitudes are simulated, where the DM signal is expected to continue growing (see Fig. 3.4) and where, for certain DM profiles, greater separation from the background emission components can be expected. Of course, the optimal search strategy depends on the DM distribution that one assumes. With distributions highly peaked near the GC, additional observations near the dynamic center of the Milky Way will likely be optimal, as opposed to the strategy pursued here. Upcoming H.E.S.S. and CTA observations of the GC will address the goal of resolving the ideal strategy for different profiles, although that is not fully addressed here. A box within $|l| < 1°$ and $|b| < 0.3°$ is masked in order to exclude sources on the Galactic plane. Furthermore, the bright source HESS J1745-303 is covered by excluding a disk of radius $0.8°$ centered at $(l, b) = (-1.29°, -0.64°)$. With this procedure, all the modeled backgrounds mentioned earlier are still to be taken care of. A mean zenith angle of $20°$ for the observations is assumed. This is an appropriate value considering the various constraints in this visibility window. Moreover, data are assumed to be collected with the full five-telescopes array CT1-5, and therefore are extracted for the corresponding instrument response functions from Holler et al. (2016). From these choices, the same energy range between 200 GeV and 70 TeV as done for the analysis presented in Chap. 7 is adopted.

A flat exposure time of 500 h distributed evenly across the inner $4°$—assumed as the ROI—of the GC is assumed. The ROI is further divided in rings of $0.1°$ width. Nevertheless, this is a conservative assumption as to the ultimate dataset H.E.S.S. can collect. Indeed, if the dataset of 546 h of observations as in Chap. 7 is considered, and one includes the dataset collected with phase I of H.E.S.S. in the inner $1°$ of the GC—also used for DM searches in Abdallah et al. (2016), Abdallah et al. (2018), one can safely claim that H.E.S.S. already has 800 h of potential ON region data near the GC. Nevertheless, the total exposure time is not distributed evenly across the inner $4°$ of the GC. For now, a flat exposure of 500 h of ON region observations will be used to compute the expected sensitivity. This makes the obtained results conservative. The limits with 1,000 h of flat exposure will also be shown later for comparison.

The analysis is assumed to exploit the conventional ON and OFF Reflected Background method as performed for Chap. 7. The IACT sensitivity is computed with the same Test-Statistics analysis that was used in Chap. 7 and the procedure follows what has been developed in Chap. 6, with the same setup which was outlined in Sect. 7.4 to compute limits on the annihilation cross section of DM particles. However, MC realizations are not performed for the computation of expected limits and uncertainty bands, but the Asimov approach is used as explained in Sect. 6.4.3.

8.3.2 Sensitivity to Two-Body Final States

This section show the limits for the model-independent approach. Limits for DM annihilating into various two-body final states are computed, assuming DM spectra determined from HDMSpectra. For the various channels considered, the sensitivity for H.E.S.S.-like observations via the computation of mean expected upper limits at 95% C.L. on the annihilation cross section $\langle \sigma v \rangle$ as a function of the DM particle mass from 0.5 up to 100 TeV is obtained (see Sect. 6.4.2).

The results are shown in the left panel of Fig. 8.5, for the non-neutrino channels and for the Einasto profile. The Figure was extracted from Montanari et al. (2023a). For a mass of 1.5 TeV, the sensitivity reaches 1.0×10^{-25} cm^{3}s^{-1} and 3.4×10^{-26} cm^{3}s^{-1} for the W^+W^- and $\tau^+\tau^-$ annihilation channels, respectively. The right panel of Fig. 8.5 shows the sensitivity for two-body neutrino final states when adopting the NFW parameterizations of the Milky Way DM distribution used in Albert et al. (2019), referred as to aNFW. For the $\nu_\tau \bar{\nu}_\tau$ channel, results assuming DM distributed according to the Einasto profile are also shown, to highlight the difference between the Einasto and the aNFW profiles. These results are compared to limits at 90% C.L. obtained with ANTARES (Albert et al. 2019) for the $\nu_\mu \bar{\nu}_\mu$ channel. For DM masses well above the weak scale, a H.E.S.S.-like IACT array is clearly sensitive to neutrino final states. This is due to the large number of VHE photons such final states can generate when electroweak corrections are incorporated. This demonstrates that IACT observations can be competitive to search for DM annihilation in these specific channels.

In order to compare these results to what was derived in Chap. 7, one can notice that, for example, for the W^+W^- channel at $m_{DM} \sim 1$TeV the expected sensitivity obtained here is weaker by a factor of ~ 2.5. This can be explained by three key points. The first and most significant difference arises from the adopted acceptances. Moreover, this work assumed 500 h observation distributed uniformly over the inner 4° against the 546 h observation of the inner 3°, and as the Einasto profile sharply peaks at inner radii, observations closer to the GC enhance the expected sensitivity for this profile. Finally, the work presented in Chap. 7 used spectra from PPPC4DMID, which predicts a slightly higher photon yield than HDMSpectra. A flat exposure of 500 h in the GC region was assumed so far. Let us now look at the results for when 1,000 h are considered as distributed evenly across the inner 4° of the GC.

As discussed in Sect. 8.3.1, 1,000 h of exposure is not too far from a realistic assumption for what should be available in the region from H.E.S.S. observations. Nevertheless, the data was accumulated during different phases of H.E.S.S. Therefore, a realistic analysis would require dedicated simulations for the instrument response functions for phase-I and phase-II data. In addition, the two datasets were collected toward slightly different regions of the GC (Abdallah et al. 2016; Abdallah et al. 2018; Abdalla et al. 2022). The currently existing data are not homogeneously spread across the considered ON region, as opposed to what is assumed now for the prospect limits. Figure 8.6 compares results obtained with 500 h and 1,000 h of flat time exposures across the considered ON region, considering DM particles annihi-

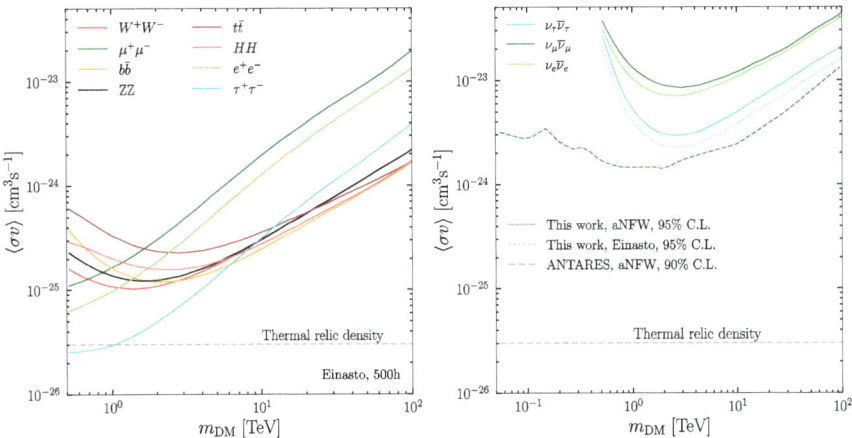

Fig. 8.5 *Left panel*: Expected upper limits on $\langle\sigma v\rangle$ as a function of the DM mass m_{DM} for six different two-body final states. All spectra are computed with `HDMSpectra`, and for the assumption of the DM distributed in the inner galaxy as the Einasto profile. The dashed gray horizontal line represents the expected cross-section for a conventional thermal relic. *Right panel*: Equivalent results for neutrino final states, and the equivalent sensitivity obtained by ANTARES for the $\nu_\mu\bar{\nu}_\mu$ channel Albert et al. (2019). To facilitate the comparison, the NFW parameters used in Albert et al. (2019) is adopted. The limits that are obtained for the $\nu_\tau\bar{\nu}_\tau$ channel are also shown for the assumption of DM distributed according to the Einasto profile, as adopted for the left panel. The Figure was extracted from Montanari et al. (2023a)

Fig. 8.6 Comparison of expected limits obtained with 500 and 1,000 h of flat time exposure. The same assumptions for the limits shown in Fig. 8.8 were made, but here the results are obtained for 500 and 1,000 h of flat time exposure across the considered ON region. The sensitivity is shown for the W^+W^- and $\tau^+\tau^-$ two-body final states. The Figure was extracted from Montanari et al. (2023a)

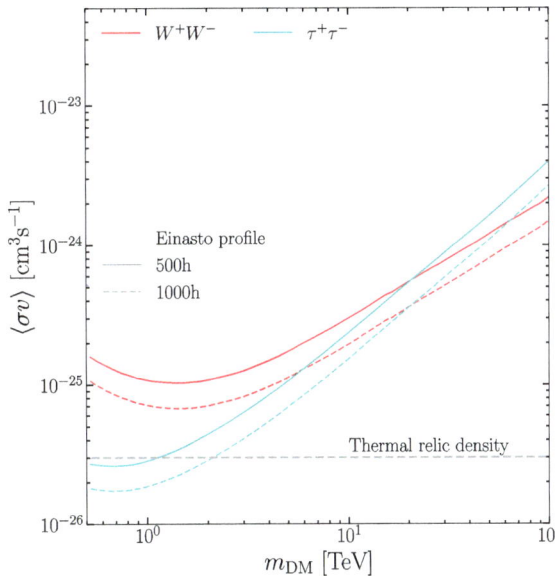

Fig. 8.7 Comparison between the Asimov and Monte Carlo simulation computations of the expected mean upper limits (solid line) and 1σ containment band (dashed line) on $\langle\sigma v\rangle$, displayed as function of m_{DM}. The limits are obtained at 95% C. L. for the W^+W^- channel derived using the computation of the gamma-ray yield from HDMSpectra and are displayed. The Figure was extracted from Montanari et al. (2023a)

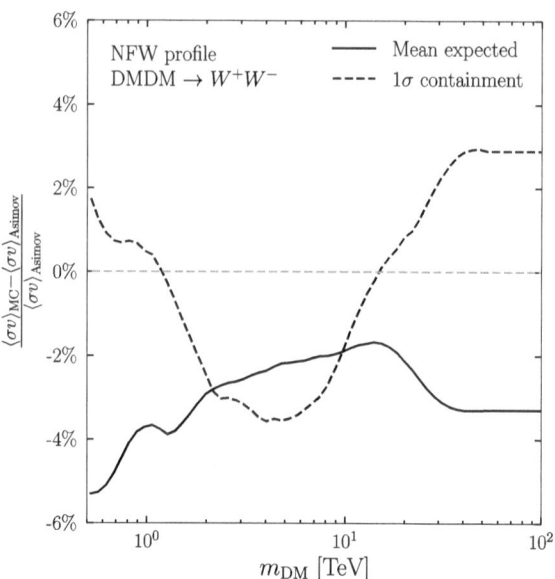

lating into the W^+W^- and $\tau^+\tau^-$ channels. The Figure was extracted from Montanari et al. (2023a).

Finally, the Asimov approach was adopted to obtain the limits shown so far in the chapter. Nevertheless, it is shown here that the Asimov approach provides an accurate determination of the sensitivity. This approach can be compared with limits derived from a Monte Carlo simulation-based approach. This comparison is shown in Fig. 8.7. The MC-based approach was built with 300 simulations. Mean expected limits and the 1σ containment band computed with the two approaches agree within 5 and 4%, respectively, in the probed mass range. The Figure was extracted from Montanari et al. (2023a).

8.3.3 Uncertainties from Theoretical Expectations

Two aspects of the DM flux in Eq. (3.7) are subject to uncertainty for a chosen mass and cross section: the spectrum shape and the spatial modeling of the DM distribution in the Milky Way. In Fig. 8.8, the impact of these uncertainties (Montanari et al. 2023a) is depicted. The results from the two possible gamm-ray yields PPPC4DMID (Cirelli et al. 2011) (blue lines) and HDMSpectra (Bauer et al. 2021) (red lines) are shown. Differences between these two methods are expected to be pronounced when m_{DM} approaches the electroweak scale, or well above it. For the lowest masses, the limits from PPPC4DMID are almost 30% stronger than HDMSpectra. The difference decreases above $\sim$1 TeV, and converge to roughly 6%. At around 100

Fig. 8.8 The systematic uncertainty on the DM signal prediction impacts on our limits, for a representative final state, the W^+W^- channel. In the top panel, the limit for the three different DM profiles are shown, the NFW (solid lines), cNFW (dashed lines), and Einasto (dotted lines). In each case, the limit obtained when using the spectrum as computed by PPPC4DMID (Cirelli et al. 2011) (blue) and HDMSpectra (Bauer et al. 2021) (red) is presented. In the lower panel, the percentage difference between the spectra for the NFW profile is highlighted

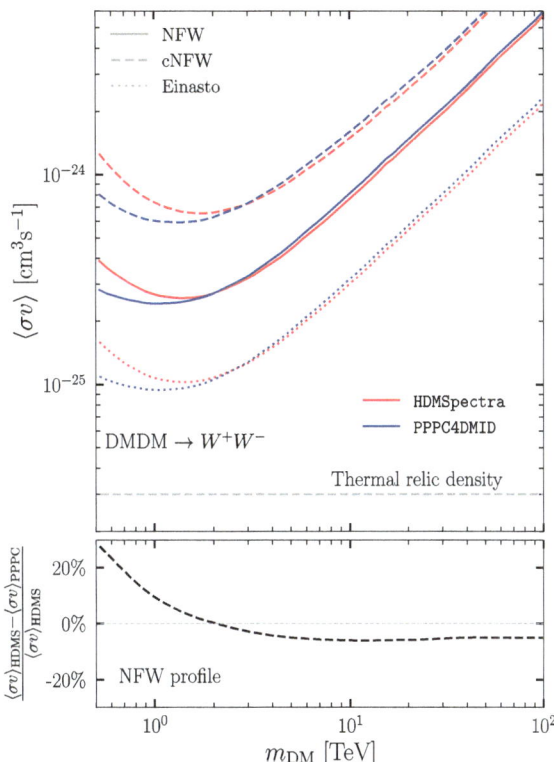

TeV, the impact of effects resulting from multiple electroweak emissions are not significant, so the two approaches are in reasonable agreement (note that HDMSpectra provides spectra for masses all the way to the Planck scale).

While the uncertainties on the gamma-ray yields are not negligible, the results in Fig. 8.8 present the dominant uncertainty in the signal prediction: the difference in the $J(\Delta\Omega)$. This reflects the present uncertainties on the DM distribution in the Milky Way. Considering the narrow FoV of H.E.S.S., this kind of analyses is still primarily sensitive to the very inner part of the Milky Way DM profile. Present observations cannot reliably probe the inner kpc, and this is evident because the cNFW leads to the weakest sensitivity at present. Adopting the cNFW over the NFW profile degrades sensitivity by a factor from 1.8 up to 3.2, depending on the mass. While challenging, any improvements in our understanding of $\rho_{DM}(r)$ in the inner galaxy will immediately result in decreased uncertainties for DM analyses.

8.3.4 Astrophysical and Instrumental Background Uncertainties

The impact of the background modeling on our search for DM signals is explored here, considering possible variations in the sources assumed as the background components.

Even with a 90% rejection efficiency, the dominant background remains hadronic cosmic-rays. Whilst the rejection efficiency can always be improved, this background appears irreducible for IACT observations. The uncertainty in the CR spectrum reaching Earth has been estimated by AMS-02 (Aguilar et al. 2015). This study established that the spectral index of the proton flux is uncertain at the level of ± 0.2, and so the central value of 2.7 is varied by this amount. This translates into an uncertainty on our limits of up to 17%, as demonstrated in Fig. 8.9. Similarly, the results were varied for the addition of an energy cutoff to the PeVatron, a change in the spectrum index of the FBs or MSP spectra at the level of ± 0.2, however in all these additional cases, no appreciable impact on the results (Montanari et al. 2023a) was found.

Figure 8.10 shows the impact of including GDE in the background contribution for our analysis (Montanari et al. 2023a). The sensitivity is computed with and

Fig. 8.9 *Top panel*: Expected upper limits on $\langle \sigma v \rangle$ as a function of the DM mass m_{DM} for the W^+W^- channel and the NFW profile parametrization. The horizontal grey long-dashed line is set to the value of the natural scale expected for the thermal-relic WIMPs. The dashed and dotted lines show the limits when the indeces of the power laws describing the spectra of cosmic rays are changed by ± 0.2. *Bottom panel*: percentage difference of the limits obtained for the two uncertainty values shown in the top panel and the limits with no uncertainty. Figure was extracted from Montanari et al. (2023a)

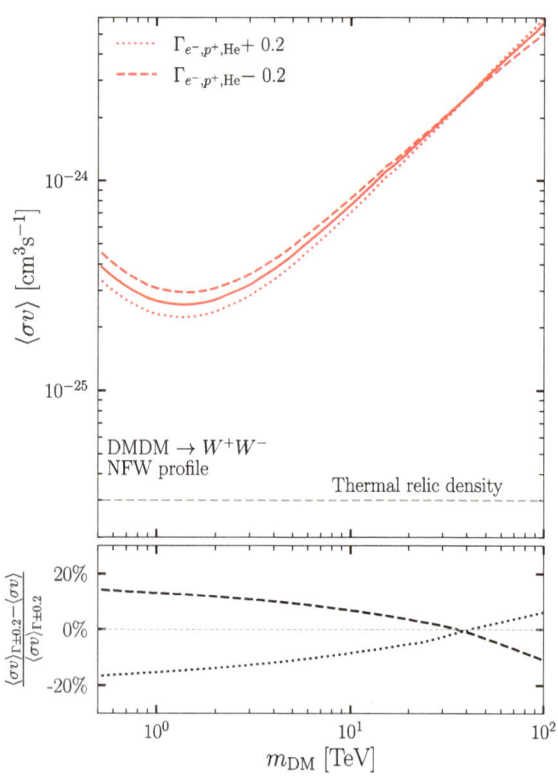

Fig. 8.10 Impact of the GDE contribution to the overall background on the expected upper limits on $\langle \sigma v \rangle$ as a function of the DM mass. The DM distribution is assumed here to follow the Einasto profile and the DM particles self-annihilate into the $W^+ W^-$ channel. Figure extracted from Montanari et al. (2023a)

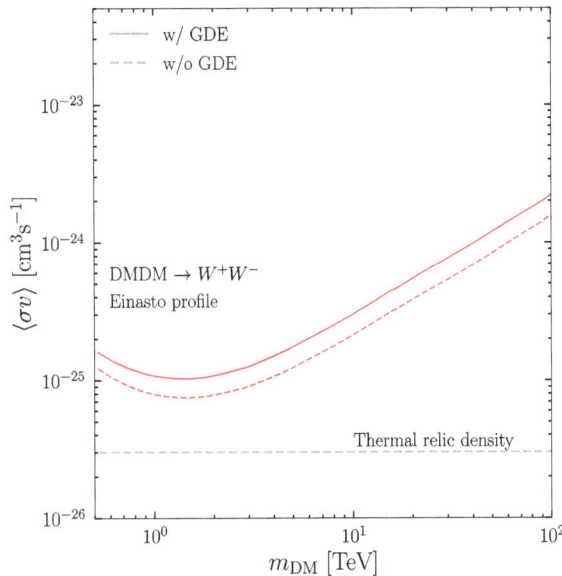

without the inclusion of the GDE in the overall background budget. Including the GDE contribution produces a sensitivity loss between 1.3 and 1.4 across the masses considered. Therefore, one can see that, with the current exposure the GDE emission starts becoming an important background emission for H.E.S.S.-like DM searches in the inner halo of the Milky Way. Such an emission could be within the reach of H.E.S.S. at TeV energies.

8.3.5 Prospect Sensitivity to Higgsino, Wino and Quintuplet Dark Matter

The procedure can be applied to canonical electroweak DM candidates of the Wino, Higgsino, and Quintuplet, as reviewed in Sect. 8.1.2.

The expected sensitivity to these scenarios is presented in Fig. 8.11 (Montanari et al. 2023a). DM distributed according to the Einasto profile has been assumed. The expected upper limits are accompanied by 1 and 2σ containment bands. In each particular case, the theoretical predictions for their cross sections are also reported. The cross-section that is considered is the weighting of the two-body photon or line final state, labeled as $\langle \sigma v \rangle_{\text{line}}$. Endpoint and continuum contributions are then weighted with respect to this cross-section. In particular, let us take $\langle \sigma v \rangle_{\text{line}} = \langle \sigma v \rangle_{\gamma\gamma} + \langle \sigma v \rangle_{\gamma Z}/2$, i.e. an appropriately weighted combination of the two-photon

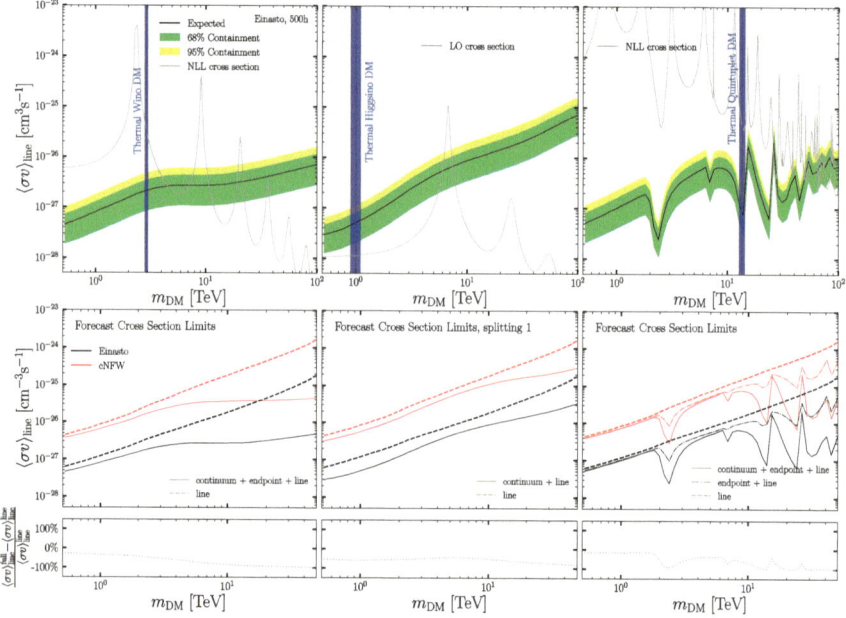

Fig. 8.11 Expected 95% C.L. upper limits on the line cross section $\langle \sigma v \rangle_{\text{line}}$ for three canonical DM models: the Wino (left), Higgsino (middle), and Quintuplet (right). The limits have been power constrained. The top panels show the sensitivity in each case, assuming an Einasto profile. The unique m_{DM} signaled out as the mass to obtain the correct relic abundance from a thermal cosmology is labeled by the vertical blue band in each case. The lower panels show results for the cNFW halo, and a breakdown of the contributions to the limits—from the line only, or the combination with the endpoint and continuum. The sharp features in the quintuplet expected limits are also explored in Fig. 8.12. Figure was extracted from Montanari et al. (2023a)

and γZ final states. The interested reader is referred to a detailed discussion of this point in Baumgart et al. (2018).

For the assumptions made in this analysis, H.E.S.S. appears sensitive to both the Wino and Quintuplet at the thermal masses. However, the predicted Higgsino flux is still out of reach of a factor of a few. Assuming all DM made of Winos, away from the thermal value, this analysis can probe masses up to 4 TeV. If the same logic is applied to the Higgsino, a small mass window near the Sommerfeld peak at 6.5 TeV can then be probed. For the Quintuplet, only certain masses above 20 TeV are out of reach.

The lower panel of Fig. 8.11, shows the expected upper limits for the halo profile and a break down of the various contributions to the spectrum. The Higgsino consists of only the line originating from the two-photon final state, in addition to continuum emission, as discussed in Sect. 8.1.2. Wino and Quintuplet spectra include the endpoint contribution, too.

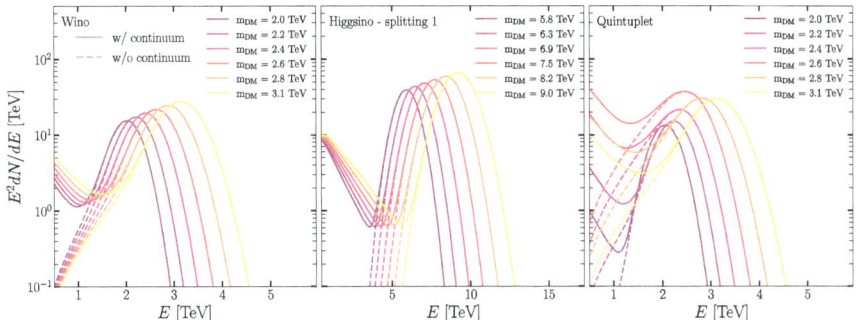

Fig. 8.12 DM annihilation spectra for the Wino (left), Higgsino (middle), and Quintuplet (right), with and without the continuum contribution (solid and dashed lines, respectively). The spectra are obtained after convolution with the H.E.S.S. energy resolution. The Wino and Quintuple spectra include the line and endpoint contributions, whereas the Higgsino includes only the line. The locations where the Quintuplet spectra evolve sharply explain the sharp variations seen in the limits as a function of mass (see Fig. 8.11). Figure extracted from Montanari et al. (2023a)

A very striking feature of the expected upper limits are the sharp features for the Quintuplet. Each one of the three DM models exhibit sharp features associated with Sommerfeld resonances for the theoretical cross-sections. Nevertheless, only the limits for the Quintuplet show these sharp properties. These features originate from sharp variations in the endpoint and continuum spectrum as a function of mass. This point is explored further in Fig. 8.12, where the spectra for three models as a function of several nearby masses (Montanari et al. 2023a) are shown. For Wino and Higgsino spectra, masses running across a Sommerfeld resonance are shown, and in both cases, the spectra vary very smoothly as a function of the mass. In the Quintuplet case, however, when one analyzes the feature's first appearance in the expected limits, it can be seen that there are large changes in the spectrum, even for very small variations in the mass. As explained in Montanari et al. (2023a), Baumgart et al. (2023), this behavior originates from the competition between the various channels that the Quintuplet can annihilate through because of the Sommerfeld process. Indeed, for the Quintuplet, the neutral initial state $\chi^0\chi^0$ can transition into a charged state $\chi^+\chi^-$, then readily annihilating into two photons. Moreover, the initial state can also transition into a doubly charged state $\chi^{++}\chi^{--}$. Both of these final states include a spectrum of Sommerfeld resonances. For the Quintuplet, for masses between $2-3$ TeV, the theoretical annihilation transitions between being dominated by the singly and doubly charged final states; one process turns off exponentially when moving away from the associated Sommerfeld peak, whereas the other turns on exponentially as the peak is approached. This causes rapid variation in the Quintuplet spectra. Multiple channels also explain the very rich Sommerfeld resonances structure in the theoretical prediction for the Quintuplet cross-section (Montanari et al. 2023a; Baumgart et al. 2023).

8.3.6   A Digression on P-Wave Annihilating Dark Matter Searches

What has been shown so far always assumed DM models with s-wave annihilation, where the annihilation cross-section of the DM particles is considered independent of the particle velocities. Instead, for velocity-dependent models, for instance, the annihilation in p-wave, the J-factor is generalized to encompass the velocity distribution of DM particles.

Indeed, one can modify Eq. (3.13) to include the cases for velocity dependent models. This extends the J-factor to include a term encompassing the DM particle velocities which assume that the DM velocity distribution $f(\vec{r}, \vec{v})$ can be written as $f(\vec{r}, \vec{v}) = \rho(\vec{r})g(\vec{v})$ normalized such that $\int d^3 v f(\vec{r}, \vec{v}) = \rho(\vec{r})$ (Jungman et al. 1996; McKeown et al. 2022). From this, the J-factor for p-wave models can then be computed. An example of J-factor profiles obtained for p-wave annihilation from the GC region has already been shown in the right panel of Fig. 3.4. The thermal $\langle \sigma v \rangle$ values obtained for the case of p-wave annihilation has been outlined in Sect. 2.1.3.

Similar to what has been done so far for s-wave annihilation, one can derive constraints on the $\langle \sigma v \rangle$ of DM particles when assuming p-wave models. Here, part of the final results obtained by the analysis presented in Montanari et al. (2023b) is shown. The authors assessed constraints on velocity-dependent models using the H.E.S.S. IGS data publicly available from H.E.S.S. Collaboration (2002). Moreover, a careful treatment of the GDE was included in the analysis, exploiting the GALPROP simulations introduced in Sect. 8.2.2. For the assumed DM distribution in the GC region, the authors adopted the J-factors as shown in Fig. 3.4, derived from FIRE-2 (which include baryon feedback processes), and DM-only simulations and extracted from McKeown et al. (2022) (see Sect. 3.3 for more details).

Figure 8.13 shows the 95% C.L. mean expected upper limits on the velocity-weighted annihilation cross section for p-wave Majorana WIMPs annihilating in the W^+W^- and $b\bar{b}$ channels (Montanari et al. 2023b). The J-factors applied in the analysis were extracted for the cases of FIRE-2 and DM-only simulations (McKeown et al. 2022). The Figure shows colored bands to present how the limits would get more/less constraining when choosing the maximum/minimum J-factors from the FIRE-2 and DM-only simulations setup. Moreover, the results show the improvement obtained with modeling of the GDE in the background for one baseline setup from the GALPROP simulations adopted in Montanari et al. (2023b). The limits reach 4.6×10^{-22} cm^{3}s^{-1} and 7.8×10^{-22} cm^{3}s^{-1} for a DM particle mass of 1.7 TeV in the W^+W^- and $b\bar{b}$ annihilation channels, respectively. For a DM mass of about 1 TeV, for annihilation in the $b\bar{b}$ channel, the limits improve upon the results derived with *Fermi*-LAT data, by the analysis in McKeown et al. (2022), by a factor $\sim$4. Although one cannot straightforwardly compare the results shown here and what was obtained with the IGS data (shown in Chap. 7) for s-waves and annihilation into the W^+W^- channel (see Fig. 7.13) considering the differences in the analyses, it is possible to make some rough estimate of what one would expect from comparing

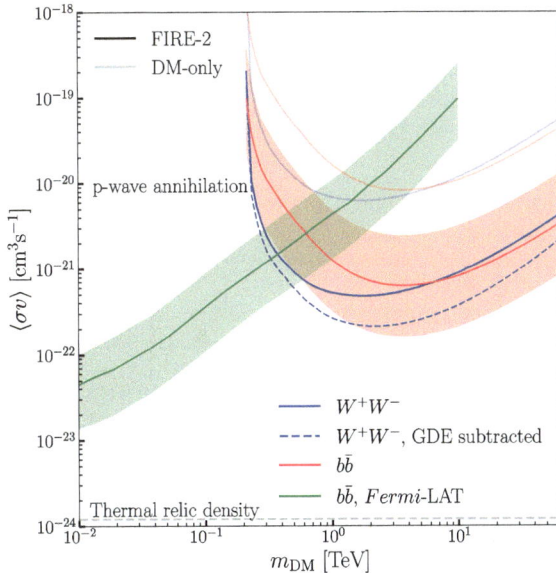

Fig. 8.13 95% C.L. mean expected upper limits on as a function of the DM mass, for p-wave annihilation scenario, in the W^+W^- (solid blue line) and $b\bar{b}$ (solid red line) channels, respectively. The limits obtained when the GDE is modeled in the background are also shown (dashed blue line). The dark and light-shaded lines correspond to limits obtained with J-factor for the FIRE-2 and DM-only simulations, respectively. The horizontal grey dashed line corresponds to the expected thermal annihilation cross section for the p-wave annihilation signal. The *Fermi*-LAT constraints are shown as a solid green line (McKeown et al. 2022). The containment bands show the limits obtained when considering the maximum and minimum values of the J-factors from the FIRE-2 simulations. Figure extracted from Montanari et al. (2023b)

the J-factor profiles. A simple estimate of the total J-factor obtained in Montanari et al. (2023b) for the region of interest inside $1°$ radius around the GC gives $J \sim 2-3 \times 10^{16}$ GeV^{2}cm^{-5}sr. Similarly, the J-factor for the Einasto profile adopted for the results shown in Fig. 7.13, sums up to $J \sim 4-5 \times 10^{21}$ GeV^{2}cm^{-5}sr. The $\sim$5 orders of magnitude difference between the integrated J-factors is roughly reflected on the difference between the limits at 1 TeV.

References

Abdalla, H., et al.: Search for dark matter annihilation signals in the H.E.S.S. inner galaxy survey. Phys. Rev. Lett. **129**(11), 111101 (2022). https://doi.org/10.1103/PhysRevLett.129.111101

Abdallah, H., et al.: Search for γ-ray line signals from dark matter annihilations in the inner galactic halo from 10 years of observations with H.E.S.S. Phys. Rev. Lett. **120**(20), 201101 (2018). https://doi.org/10.1103/PhysRevLett.120.201101

Abdallah, H., et al.: Search for dark matter annihilations towards the inner Galactic halo from 10 years of observations with H.E.S.S. Phys. Rev. Lett. **117**(11), 111301 (2016). https://doi.org/10. 1103/PhysRevLett.117.111301

Abdollahi, S., et al.: Search for cosmic-ray electron and positron anisotropies with seven years of fermi large area telescope data. Phys. Rev. Lett. **118**(9), 091103 (2017). https://doi.org/10.1103/ PhysRevLett.118.091103

Abramowski, A., et al.: Acceleration of petaelectronvolt protons in the galactic centre. Nature **531**, 476 (2016). https://doi.org/10.1038/nature17147

Ackermann, M., et al.: Fermi-LAT observations of the diffuse γ-ray emission: implications for cosmic rays and the interstellar medium. Astrophys. J. **750**(1), 3 (2012). https://doi.org/10.1088/ 0004-637x/750/1/3, ISSN: 1538-4357

Ackermann, M., et al.: The fermi Galactic center GeV excess and implications for dark matter. Astrophys. J. **840**(1), 43 (2017). https://doi.org/10.3847/1538-4357/aa6cab

Aguilar, M., et al.: Precision measurement of the proton flux in primary cosmic rays from rigidity 1 GV to 1.8 TV with the alpha magnetic spectrometer on the international space station. Phys. Rev. Lett. **114**, 171103 (2015). https://doi.org/10.1103/PhysRevLett.114.171103

Ajello, M., et al.: Fermi-LAT observations of high-energy γ-ray emission toward the galactic center. Astrophys. J. **819**(1), 44 (2016). https://doi.org/10.3847/0004-637X/819/1/44

Albert, A., et al.: Results from the search for dark matter in the Milky Way with 9 years of data of the ANTARES neutrino telescope. Phys. Lett. B **769**, 249–254 (2017). https://doi.org/10.1016/j. physletb.2017.03.063. [Erratum: Phys. Lett. B 796, 253–255 (2019)]

Arkani-Hamed, N., et al.: Simply unnatural supersymmetry (2012). arXiv: 1212.6971 [hep-ph]

Bauer, C.W., Rodd, N.L., Webbe, B.R.: Dark matter spectra from the electroweak to the planck scale. JHEP **06**, 121 (2021). https://doi.org/10.1007/JHEP06(2021)121

Baumgart, M., et al.: Resummed photon spectra for WIMP annihilation. JHEP **03**, 117 (2018). https://doi.org/10.1007/JHEP03(2018)117, arXiv: 1712.07656 [hep-ph]

Baumgart, M., et al.: The quintuplet annihilation spectrum (2023). https://doi.org/10.48550/arXiv. 2309.11562, arXiv: 2309.11562 [hep-ph]

Baumgart, M., Cohen, T., Moulin, E., et al.: Precision photon spectra for Wino annihilation. JHEP **01**, 036 (2019). https://doi.org/10.1007/JHEP01(2019)036

Beneke, M., et al.: Relic density of wino-like dark matter in the MSSM. JHEP **03**, 119 (2016). https://doi.org/10.1007/JHEP03(2016)119, arXiv: 1601.04718 [hep-ph]

Beneke, M., et al.: Precise yield of high-energy photons from Higgsino dark matter annihilation. J. High Energy Phys. **2020**(3), 30 (2020). https://doi.org/10.1007/JHEP03(2020)030

Bernlöhr, K., et al.: Monte Carlo design studies for the Cherenkov telescope array. Astropart. Phys. **43**, 171–188 (2013). https://doi.org/10.1016/j.astropartphys.2012.10.002, arXiv: 1210.3503 [astro-ph.IM]

Bernlohr, K.: Simulation of imaging atmospheric Cherenkov telescopes with CORSIKA and sim_telarray. Astropart. Phys. **30**, 149–158 (2008). https://doi.org/10.1016/j.astropartphys.2008. 07.009

Bottaro, S., et al.: The last complex WIMPs standing (2022b). arXiv: 2205.04486 [hep-ph]

Bottaro, S., et al.: Closing the window on WIMP dark matter. Eur. Phys. J. C **82**(1), 31 (2022a). https://doi.org/10.1140/epjc/s10052-021-09917-9

Cautun, M., et al.: The Milky Way total mass profile as inferred from Gaia DR2. Mon. Not. Roy. Astron. Soc. **494**(3), 4291–4313 (2020). https://doi.org/10.1093/mnras/staa1017

Cerutti, B., Philippov, A.A., Spitkovsky, A.: Modelling high-energy pulsar light curves from first principles. Mon. Not. Roy. Astron. Soc. **457**(3), 2401–2414 (2016). https://doi.org/10.1093/ mnras/stw124

Cirelli, M., et al.: Gamma ray tests of minimal dark matter. JCAP **1510**(10), 026 (2015). https:// doi.org/10.1088/1475-7516/2015/10/026, arXiv: 1507.05519 [hep-ph]

Cirelli, M., et al.: PPPC 4 DM ID: a poor particle physicist cookbook for dark matter indirect detection. JCAP **1103**, 051 (2011). https://doi.org/10.1088/1475-7516/2012/10/E01, https://doi. org/10.1088/1475-7516/2011/03/051

Cirelli, M., Fornengo, N., Strumia, A.: Minimal dark matter. Nucl. Phys. **B753**, 178–194 (2006). https://doi.org/10.1016/j.nuclphysb.2006.07.012, arXiv: hep-ph/0512090 [hep-ph]

Cirelli, M., Franceschini, R., Strumia, A.: Minimal dark matter predictions for galactic positrons, anti-protons, photons. Nucl. Phys. **B800**, 204–220 (2008). https://doi.org/10.1016/j.nuclphysb. 2008.03.013, arXiv: 0802.3378 [hep-ph]

Cirelli, M., Strumia, A., Tamburini, M.: Cosmology and astrophysics of minimal dark matter. Nucl. Phys. **B787**, 52–175 (2007). https://doi.org/10.1016/j.nuclphysb.2007.07.023, arXiv: 0706.4071 [hep-ph]

Cirelli, M., Strumia, A.: Minimal dark matter: model and results. New J. Phys. **11**, 105005 (2009). https://doi.org/10.1088/1367-2630/11/10/105005, arXiv: 0903.3381 [hep-ph]

Co, R.T., et al.: Discovery potential for split supersymmetry with thermal dark matter (2022b). arXiv: 2206.11912 [hep-ph]

Co, R.T., Sheff, B., Wells, J.D.: Race to find split Higgsino dark matter. Phys. Rev. D **105**(3), 035012 (2022a). https://doi.org/10.1103/PhysRevD.105.035012

H.E.S.S. Collaboration (2002). https://www.mpi-hd.mpg.de/hfm/HESS/pages/about/

de Naurois, M., Rolland, L.: A high performance likelihood reconstruction of γ-rays for imaging atmospheric Cherenkov telescopes. Astropart. Phys. **32**, 231–252 (2009). https://doi.org/10.1016/ j.astropartphys.2009.09.001, arXiv: 0907.2610 [astro-ph.IM]

Fox, P.J., Kribs, G.D., Adam M.: Split Dirac supersymmetry: an ultraviolet completion of Higgsino dark matter. Phys. Rev. **D90**(7), 075006 (2014). https://doi.org/10.1103/PhysRevD.90.075006

Goodenough, L., Hooper, D.: Possible evidence for dark matter annihilation in the inner milky way from the fermi gamma ray space telescope (2009). arXiv: 0910.2998 [hep-ph]

Guépin, C., Cerutti, B., Kotera, K.: Proton acceleration in pulsar magnetospheres. Astron. Astrophys. **635**, A138 (2020). https://doi.org/10.1051/0004-6361/201936816

Hall, L.J., Nomura, Y., Shirai, S.: Spread supersymmetry with wino LSP: Gluino and dark matter signals. JHEP **01**, 036 (2013). https://doi.org/10.1007/JHEP01(2013)036

Herold, L., Malyshev, D.: Hard and bright gamma-ray emission at the base of the fermi bubbles. Astron. Astrophys. **625**, A110 (2019). https://doi.org/10.1051/0004-6361/201834670

Hisano, J., et al.: Non-perturbative effect on thermal relic abundance of dark matter. Phys. Lett. B **646**, 34–38 (2007). https://doi.org/10.1016/j.physletb.2007.01.012, arXiv: hep-ph/0610249 [hep-ph]

Holler, M., et al.: Photon reconstruction for H.E.S.S. using a semi-analytical model. PoS ICRC2015, 980 (2016). https://doi.org/10.22323/1.236.0980

Hooper, D., Goodenough, L.: Dark matter annihilation in the galactic center as seen by the fermi gamma ray space telescope. Phys. Lett. B **697**, 412–428 (2011). https://doi.org/10.1016/j. physletb.2011.02.029

Hryczuk, A., Iengo, R., Ullio, P.: Relic densities including Sommerfeld enhancements in the MSSM. JHEP **03**, 069 (2011). https://doi.org/10.1007/JHEP03(2011)069, arXiv: 1010.2172 [hep-ph]

Jungman, G., Kamionkowski, M., Griest, K.: Supersymmetric dark matter. Phys. Rept. **267**, 195–373 (1996). https://doi.org/10.1016/0370-1573(95)00058-5

Kearney, J., Orlofsky, N., Pierce, A.: Z boson mediated dark matter beyond the effective theory. Phys. Rev. D **95**(3), 035020 (2017). https://doi.org/10.1103/PhysRevD.95.035020

Leane, R.K., et al.: Snowmass 2021 cosmic frontier white paper: puzzling excesses in dark matter searches and how to resolve them (2022). arXiv: 2203.06859 [hep-ph]

Lefranc, V., et al.: Prospects for annihilating dark matter in the inner Galactic halo by the Cherenkov telescope array. Phys. Rev. D **91**(12), 122003 (2015). https://doi.org/10.1103/PhysRevD.91. 122003

Macias, O., et al.: Cherenkov telescope Array sensitivity to the putative millisecond pulsar population responsible for the galactic centre excess. Mon. Not. Roy. Astron. Soc. **506**(2), 1741–1760 (2021). https://doi.org/10.1093/mnras/stab1450, arXiv: 2102.05648 [astro-ph.HE]

Mahbubani, R., Senatore, L.: The minimal model for dark matter and unification. Phys. Rev. **D73**, 043510 (2006). https://doi.org/10.1103/PhysRevD.73.043510

McKeown, D., et al.: Amplified J-factors in the galactic centre for velocity-dependent dark matter annihilation in FIRE simulations. Mon. Not. Roy. Astron. Soc. **513**(1), 55–70 (2022). https://doi.org/10.1093/mnras/stac966

Mitridate, A., et al.: Cosmological implications of dark matter bound states. JCAP **1705**(05), 006 (2017). https://doi.org/10.1088/1475-7516/2017/05/006, arXiv: 1702.01141 [hep-ph]

Montanari, A., Macias, O., Moulin, E.: TeV gamma-ray sensitivity to velocity-dependent dark matter models in the Galactic center. Phys. Rev. D **108**(8), 083027 (2023b). https://doi.org/10.1103/PhysRevD.108.083027, https://link.aps.org/doi/10.1103/PhysRevD.108.083027

Montanari, A., Moulin, E., Rodd, N.L.: Toward the ultimate reach of current imaging atmospheric Cherenkov telescopes and their sensitivity to TeV dark matter". Phys. Rev. D **107**(4), 043028 (2023). https://doi.org/10.1103/PhysRevD.107.043028, arXiv: 2210.03140 [astro-ph.HE]

Montanari, A., Moulin, E., Rodd, N.L.: Toward the ultimate reach of current imaging atmospheric Cherenkov telescopes and their sensitivity to TeV dark matter. Phys. Rev. D **107**(4), 043028 (2023a). https://doi.org/10.1103/PhysRevD.107.043028, https://link.aps.org/doi/10.1103/PhysRevD.107.043028

Moulin, E., et al.: Search for TeV emission from the fermi bubbles at low galactic latitudes with H.E.S.S. inner galaxy survey observations. PoS ICRC2021, 791 (2021). arXiv: 2108.10028 [astro-ph.HE]

Moulin, E.: The inner 300 parsecs of the milky way seen by H.E.S.S.: a Pevatron in the Galactic centre. In: EPJ Web Conferences, vol. 136, p. 03017 (2017). https://doi.org/10.1051/epjconf/201713603017

Porter, T.A., Jóhannesson, G., Moskalenko, I.V.: The GALPROP cosmic-ray propagation and non-thermal emissions framework: release v57. Astrophys. J. Suppl. Ser. **262**(1), 30 (2022). https://doi.org/10.3847/1538-4365/ac80f6

Queiroz, F.S., Yaguna, C.E., Weniger, C.: Gamma-ray limits on neutrino lines. JCAP **05**, 050 (2016). https://doi.org/10.1088/1475-7516/2016/05/050

Rinchiuso, L., et al.: Prospects for detecting heavy WIMP dark matter with the Cherenkov telescope array: the Wino and Higgsino. Phys. Rev. D **103**(2), 023011 (2021). https://doi.org/10.1103/PhysRevD.103.023011

Schlegel, D.J., Finkbeiner, D.P., Davis, M.: Maps of dust IR emission for use in estimation of reddening and CMBR foregrounds. Astrophys. J. **500**, 525 (1998). https://doi.org/10.1086/305772

Silverwood, H., et al.: A realistic assessment of the CTA sensitivity to dark matter annihilation. JCAP **03**, 055 (2015). https://doi.org/10.1088/1475-7516/2015/03/055

Springel, V., White, S.D.M., Frenk, C.S., et al.: A blueprint for detecting supersymmetric dark matter in the Galactic halo. Nature **456**, 73–76 (2008)

Storm, E., Weniger, C., Calore, F.: SkyFACT: high-dimensional modeling of gamma-ray emission with adaptive templates and penalized likelihoods. JCAP **1708**(08), 022 (2017). https://doi.org/10.1088/1475-7516/2017/08/022

Su, M., Slatyer, T.R., Finkbeiner, D.P.: Giant gamma-ray bubbles from fermi-LAT: active galactic nucleus activity or bipolar galactic wind? Astrophys. J. **724**(2), 1044–1082 (2010). https://doi.org/10.1088/0004-637X/724/2/1044

Chapter 9
Outlook

This book aims at defining the framework to carry out a search for dark matter (DM) annihilation signals, and presenting the achievable sensitivity of the current generation of Imaging Atmospheric Cherenkov Telescopes (IACTs) to annihilating dark matter. The already obtained constraints and the near-future forecast sensitivity are shown. A considerable range of possible dark matter annihilation models is explored, including a broad choice of two-body final states and canonical TeV electroweak candidates—arising as specific Weakly Interacting Massive Particles scenario realizations, for DM masses between a few hundred GeV up to $\sim$100 TeV. State-of-the-art computations for the DM spectra and models for the DM density distribution in the inner Galaxy are presented.

It is demonstrated that observations with H.E.S.S.-like IACTs are sensitive enough to start probing the thermal relic prediction for a range of two-body channels for $m_{\mathrm{DM}} \sim 1$ TeV. It is also highlighted that IACT observations can compete with neutrino telescopes—such as ANTARES—to constrain TeV neutrino line channels; these effects imply that searching for heavier DM is inherently multimessenger. For specific DM particle candidates actively searched in direct detection and collider experiments, H.E.S.S.-like instruments reach the required sensitivity to probe the thermal Wino and Quintuplet. Despite all, the thermal Higgsino may be currently still out of reach by a factor of a few, but within the detection range in the near future with current and future facilities. For the results presented in this book, about 500 h of exposure in the Galactic center (GC) region, being either measured or simulated were utilized. A more comprehensive look at the total time available with H.E.S.S. results in about 800 h including all the phases of data taking with the array. At the moment of the writing, H.E.S.S. is continuously collecting data in the GC region, and a total of more than 1,000 h of exposure should be achieved within a couple of years. All of this, summed together, could push the reach even further. The limits obtained with the observations for the Inner Galaxy Survey program provide the strongest constraints

© The Author(s), under exclusive license to Springer Nature Switzerland AG 2024
A. Montanari and E. Moulin, *Searching for Dark Matter with Imaging Atmospheric Cherenkov Telescopes*, https://doi.org/10.1007/978-3-031-66470-0_9

so far at the TeV energy range. The next observatory in the field, the Cherenkov Telescope Array (CTA), should collect even larger datasets.

While tremendous developments are being deployed to improve the determination of the DM distribution in the inner Galaxy, its limited knowledge induces a systematic uncertainty in IACT DM searches. This uncertainty is largest near the dynamic center of the Galaxy. Further improvements in its determination will immediately translate into lower systematic uncertainties. Another limitation remains the statistical uncertainty, highlighting the importance of continued data collection with current-generation IACTs. The impact of the uncertainties from the various background sources and modeling has been shown to be presently negligible compared to the statistical ones.

Alternative techniques for background determination beyond the standard methods can also be pursued. Several efforts are currently deployed for the development of background models. Nevertheless, such approaches need large amounts of data taken in empty, e.g., extragalactic, fields and still have to reach the level of control of the systematic uncertainties required to be effectively used in the GC region. Since IACTs like H.E.S.S. continue to collect GC observations, it will become important to study the optimal survey strategies to balance both the DM sensitivity reach and the systematic robustness of the results.

Even after more than a decade of searches for DM annihilation signals in the TeV energy range pursued in a variety of astrophysical environments, there are still parameter spaces to explore for simple DM models not yet detected. While CTA will soon begin observing and collecting an unprecedented amount of data for potentially revolutionary results, a careful assessment of systematic uncertainties is mandatory, with its highest possible control, in order to achieve further significant improvements with respect to the currently operating IACTs. In the lively field of DM searches, the first hints of DM annihilations may well emerge in the near future.